Our Backyard

Our Backyard

A Quest for Environmental Justice

Edited by
Gerald R. Visgilio
Diana M. Whitelaw

ROWMAN & LITTLEFIELD PUBLISHERS, INC.
Lanham • Boulder • New York • Oxford

GE
230
.O77
2003
June 2004

ROWMAN & LITTLEFIELD PUBLISHERS, INC.

Published in the United States of America
by Rowman & Littlefield Publishers, Inc.
A Member of the Rowman & Littlefield Publishing Group
4501 Forbes Boulevard, Suite 200, Lanham, Maryland 20706
www.rowmanlittlefield.com

PO Box 317
Oxford
OX2 9RU, UK

Copyright © 2003 by Rowman & Littlefield Publishers, Inc.

British Library Cataloguing in Publication Information Available

Library of Congress Cataloging-in-Publication Data

Our backyard : a quest for environmental justice / edited by Gerald R.
Visgilio and Diana M. Whitelaw.
    p. cm.
Includes bibliographical references and index.
    ISBN 0-7425-2362-4 (hardcover : alk. paper) — ISBN 0-7425-2363-2
(pbk. : alk. paper)
    1. Environmental justice—United States. I. Visgilio, Gerald, 1941–
II. Whitelaw, Diana, 1943–
    GE230 .O77 2003
    363.7'0525—dc21

                                                          2002152025

Printed in the United States of America

∞™ The paper used in this publication meets the minimum requirements of
American National Standard for Information Sciences—Permanence of Paper
for Printed Library Materials, ANSI/NISO Z39.48-1992.

Contents

Preface

On April 20 and 21, 2001, Connecticut College presented a conference, "A Quest for Environmental Justice: Healthy, High Quality Environments for all Communities." Two grassroots organizations, the Connecticut Coalition for Environmental Justice and the Southeastern Connecticut Indoor Air Quality Coalition, joined the college's Goodwin-Niering Center for Conservation Biology and Environmental Studies as conference sponsors. The overall goal of the conference was to explore whether minorities and poor communities are victims of environmental injustices. The Connecticut College conference provided an opportunity for leading scholars to discuss the history, status, and dilemmas of the environmental justice movement. Bunyan Bryant, the director of the Environmental Justice Initiative at the University of Michigan, gave the keynote address. Conference speakers focused on topics that included the history of the environmental justice movement in the United States, green imperialism, environmental justice and the social determinants of health, empirical research and case studies, alternative approaches to confronting environmental injustices, and the political sustainability of the environmental justice movement. The audience included representatives of federal, state, and local agencies, concerned citizens, individuals from nongovernmental organizations, and students and faculty from Connecticut College and other universities. This book is based on the papers presented at this conference.

Introduction

Gerald R. Visgilio and Diana M. Whitelaw

Do historically disadvantaged communities incur a disproportionate share of society's environmental risks? Do these risks result in significant and widespread health problems for racial minorities and the poor? Should activists and policy makers seek policy remedies for all environmental inequities? Environmental justice is a grassroots movement that deals with environmental burdens and their distributional consequences. It is a form of community empowerment—a desire by minorities and the poor to actively participate in the decision-making process as it pertains to local environmental issues. Grassroots activists in the environmental justice movement are concerned citizens (often women) who see "their families, homes, and communities threatened" by toxic waste and other pollutants.[1] In advancing its agenda of social equity, the environmental justice movement takes the position that historically disadvantaged communities are more likely than others to be damaged by pollution and less likely to benefit from the enforcement of environmental regulations. This is a position that provokes an adversarial cry of environmental racism, a cry that is perceived to be a "superb mobilizing tool" in the "repertoire of community advocacy."[2] In this respect, environmental justice advocacy draws attention to the "inescapable" distributional effect that "what some people do with environmental resources has great impact on the welfare" of other individuals in our society.[3]

In 1982, environmental justice became a national issue when several hundred individuals protested the siting of a landfill in Warren County, North Carolina. The Warren County demonstration has been described as the "first national protest by blacks on the hazardous waste issue."[4] The protest prompted the General Accounting Office (GAO) to examine the siting of hazardous waste landfills in southern states. In a 1983 report, the GAO found that three of the four offsite hazardous waste landfills in the Environmental Protection Agency's (EPA's) Region IV were located in communities where the local population was predominantly black and poor.[5] The United Church of Christ Commission for Racial Justice looked at the siting issue at the national level. Their 1987 report shows race to be "most significant among variables tested" in the siting of hazardous waste facilities.[6] Pressure from the environmental justice movement elicited a policy response at the federal level. In November 1992, the Office of Environmental Equity was created within the EPA. In February 1994, President Clinton issued Executive Order 12898, which required the inclusion of environmental justice in federal policy.

Many empirical studies support the position that minorities and the poor are being unfairly burdened with environmental risks. In addition to showing that locally undesirable land uses (LULUs) are more likely to be placed in economically depressed or minority communities, these studies also deal with issues such as the health problems from lead paint exposure for minority children and the occupational risks of chemical exposure for migrant farmworkers. Critics of the environmental justice movement contend that studies showing an unfair environmental burden often suffer from evidentiary and methodological shortcomings. They also argue that other empirical evidence fails to support the position of an overrepresentation of minorities and the poor in communities hosting hazardous facilities. A 1994 study in *Demography* by D. L. Anderton et al., for example, found that racial minorities were not disproportionately located in census tracts that housed commercially hazardous waste facilities.[7]

Effective citizen involvement in local or neighborhood issues is fundamental to the success of the environmental justice movement. Health is an issue in which environmental justice activists mobilize community support against alleged inequalities. Minority and poor communities are indeed blighted with environmental burdens, burdens that some environmental justice advocates describe as quality-of-life issues such as traffic congestion, noxious odors, noise, dilapidated buildings, and unsanitary living conditions. Health in the broader context of quality of life may provide a more accurate description of the aspirations of individuals who fight against environmental and other injustices. Inequality of opportunity also afflicts minority and poor communities. Because of limited resources, individuals in these communities consume relatively little of society's goods and services. Enhanced economic opportunity for disadvantaged people is an important aspect in the battle for equality. Brownfields redevelopment projects are often seen as a way to raise the living standard in poor and minority communities. In this respect, environmental justice activism provides an opportunity for minorities and the poor to participate in the discussion pertaining to the tradeoff between economic growth and environmental risk.[8]

Our Backyard: A Quest for Environmental Justice presents the views of scholars and activists with respect to the history, status, and dilemmas of environmental justice. *Our Backyard* is an interdisciplinary approach to the topic of environmental justice. In this respect, we believe that our book is a stimulating collaboration of positions reflecting a wide range of academic disciplines such as ethnobotany, philosophy, political science, law, medicine, and sociology. In addition to contributions by academicians, we have also included the views of activists, individuals from nongovernmental organizations (NGOs), and representatives of federal and state government. Because *Our Backyard* deals with a wide array of environmental justice issues, it should be of interest to a broad audience that would include academicians, researchers, activists, advocacy groups, policy makers, and concerned citizens. *Our Backyard* also may be used as a book of readings for undergraduate and beginning graduate courses in environmental justice, environmental policy, environmental studies, or criminal justice. The book raises important questions that will engender considerable debate about environmental justice for many years. We believe the following are examples of questions that should provide fertile ground for dialogue between activists and policy makers or between teachers and students: Are poor and minority communities unfairly exposed to environmental hazards? If so, do these hazards create serious health risks for racial minorities and the poor? Does the existing empirical evidence support the position of widespread environmental injustices? Is the major research issue discriminatory siting or disparate risk? How can disadvantaged communities more effectively protect their neighborhoods? Does wealth redistribution necessarily underlie the achievement of environmental equality? What is the role of women in the environmental justice movement? How should society's scarce resources be allocated when addressing quality-of-life issues in disadvantaged communities? Is the environmental justice movement politically sustainable in a more conservative political climate? How should the issue of green imperialism be addressed?

ENVIRONMENTAL HAZARDS IN
POOR AND MINORITY COMMUNITIES

The purpose of part I of this book is to provide an overview of the history of the environmental justice movement and to introduce important issues and challenges faced by poor and minority communities. In this section of the book, we also introduce a philosophical perspective on quality-of-life issues, along with the argument that a more egalitarian society is a prerequisite for achieving environmental justice in a broader dimension. We then explore the issue of environmental justice and resource conservation in the context of a developing country. In chapter 1, "History and Issues of the Environmental Justice Movement," Bunyan Bryant discusses the struggles of the environmental justice movement in the United States. Bryant casts the history of the environmental justice movement in terms of the civil rights movement and other social events of the 1960s. He recognizes the importance of several conferences that advanced the cause of environmental justice. According to Bryant, the 1976 conference held at Black Lake, Michigan, was the "precursor" of other important conferences. In addition to discussing these conferences, Bryant presents an overview of the legislative history pertaining to environmental justice. Bryant also recognizes that the dumping of contaminated soil in Warren County provided "activists with an opportunity for championing civil rights" in a different light. By uniting civil rights and environmental activists, the Warren County demonstration gave rise to "the contemporary environmental justice movement." With respect to the national debate about environmental justice, Bryant discusses methodological differences among several studies that deal with race and the siting of hazardous waste facilities. For him, the findings of these studies are important because of their impact on government policy, public opinion, and minority communities. Bryant concludes by discussing the backlash against the environmental justice movement during the late 1990s.

In chapter 2, "Environmental Justice and the Social Determinants of Health," Virginia Ashby Sharpe uses empirical evidence to discuss the relationship between the social determinants of public health and the concept of environmental justice. In pursuing the question of what constitutes a good society, Sharpe argues that environmental justice literature should move beyond its narrow focus on discriminatory siting and disparate risk. She notes that the distribution of income significantly impacts social cohesion, which in turn is an underlying determinant of health. According to Sharpe, health, when viewed in the more inclusive framework of "public health and development paradigms," provides a deeper understanding of what it means for an individual or a society to be healthy. She sees policies that redistribute wealth as fundamental to achieving environmental justice in a broader context. Sharpe also examines the meaning of "treating unequals unequally" in the pursuit of environmental justice.

Ethnobotanist Manuel Lizarralde brings a global perspective to the book with his discussion of green imperialism. In chapter 3, "Green Imperialism: Indigenous People and Conservation of Natural Environments," Lizarralde writes that green imperialism is often based on the unrealistic view of the relationship between indigenous people and developed societies. He questions the idea of portraying indigenous people as "Noble Savages" who must protect their environment for the benefit of developed countries. Lizarralde focuses on the Bari people, an indigenous society in Venezuela, as an example of how green imperialism adversely affects indigenous people and developing nations. Lizarralde argues that multinational corporations use "biodiversity prospecting" to exploit indigenous peoples by taking raw materials from their lands without just compensation. These resources include genetic materials from which the agricultural and pharmaceutical enterprises realize enormous profits.

EMPIRICAL RESEARCH AND METHODOLOGICAL CHALLENGES

The growing body of empirical research generated by concerns for environmental justice has produced contradictory results and inconsistent conclusions. In part II, we present a review of environmental equity studies and an example of research on the impact of regional systems for solid waste disposal, using multivariate techniques to determine whether race or poverty are predictors of distance from hazards. We then provide the reader with some practical and methodological limitations in the field of environmental justice research. In chapter 4, "Burning and Burying in Connecticut: Are Regional Solutions to Solid Waste Disposal Equitable?" Timothy Black and John Stewart focus on Connecticut's transition to a regional system for solid waste disposal. In response to federal regulations, states throughout the country are replacing "old town dumps" with regional disposal systems. Regional systems, because they concentrate the disposal process, often engender considerable debate. State officials argue that newer disposal systems are less damaging to the environment, while neighborhood groups complain that regional systems transform minority communities into dumping grounds for solid waste. Black and Stewart contend that Connecticut, which is in the process of developing a regional system, provides a "unique opportunity for research to inform public policy." Black and Stewart examine the siting of regional facilities in Connecticut from an environmental equity perspective. In their analysis, they show that the percentage of minorities in the surrounding area is an important determinant in the location of a regional disposal facility.

In chapter 5, "Risky Business? Relying on Empirical Studies to Assess Environmental Justice," Pamela Davidson characterizes the use of empirical evidence to assess environmental injustices as "risky business." Discussing the methodological challenges facing environmental justice research, Davidson looks at the "lack of clarity" concerning the research question. Does the major research question involve discriminatory siting or disparate risk? Environmental justice research also is challenged by the choice of a unit of analysis. In environmental justice studies, the unit of analysis has ranged from zip codes or counties to census tracts or block areas. Here, Davidson argues that the use of different units of analysis is an underlying explanation of contradictory research results. She also contends that the lack of clarity with respect to the research hypothesis may result in the use of an inappropriate unit of analysis. Davidson notes that studying the effects of one toxin at a time fails to capture the synergistic effects of chemical substances. She also questions the validity of many recent studies and suggests new steps be taken to improve future research.

RESPONSES TO ENVIRONMENTAL INJUSTICES

Environmental injustice provokes a wide range of responses, from anger and despair, to political activity, to increased education and involvement of activist citizens and community groups. In part III, we offer an overview of political responses followed by case studies of collective mobilization in communities confronting injustices. These cases contrast strategies used by a regional coalition focused on community education and advocacy with those used by an individual activist mobilizing her community; also, the cases include the gender-based views of indigenous and Hispanic women leaders who are community activists or public officials. In chapter 6, "Syndrome Behavior and the Politics of Environmental Justice," Harvey White addresses the issue of syndrome behaviors associated with environmental justice. Syndrome behavior involves actions taken by communities and politicians in response to actual or perceived environmental risks. White discusses several types of syndrome behaviors, such as Not In My Back Yard (NIMBY), Not in My Election Year (NIMEY), Not in My Term of Office (NIM-TOO), Put It In Their Back Yard (PIITBY), and Why In My Back Yard? (WIMBY). He argues that the NIMBY syndrome is more likely to occur in economically and politically elite communities, while WIMBY is generally found in less-affluent and minority communities. From White's perspective, poor minority communities often lack the resources to wage a successful NIMBY campaign and, as a consequence, they tend to be more "reactive than proactive" in their responses to environmental risks. White concludes his chapter with a discussion of alternative views of the politics of environmental justice.

In chapter 7, "Confronting Environmental Injustice in Connecticut," the authors focus on ways in which activism, education, and involvement are important when ordinary citizens organize to fight against environmental injustices in their communities. They discuss several approaches to initiating change that are designed to protect low-income communities and communities of color from environmental injustices. James Younger describes how EPA New England incorporates community-based environmental protection programs to address environmental injustices. He focuses on an innovative regional EPA pilot program, the Urban Environmental Initiative, designed to tackle urban environmental and associated public health problems. Cynthia

Jennings describes how one person can make a difference by mobilizing the community across racial and cultural lines. She describes her personal involvement in the community fight against a proposed landfill expansion in Hartford, Connecticut. At the end of the chapter, Mark Mitchell describes how, through education and advocacy, the Connecticut Coalition for Environmental Justice has improved the environment, health, and quality of life of urban residents. He stresses the importance of educating and organizing local communities when attempting to affect changes in discriminatory state policies for siting environmentally risky facilities.

In chapter 8, "For the People: American Indian and Hispanic Women in New Mexico's Environmental Justice Movement," Diane-Michele Prindeville discusses the role of women in their fight against environmental injustices in New Mexico. Prindeville argues that the goals of indigenous and Hispanic women are rooted in issues of social equity and environmental justice. Furthermore, her research shows that indigenous and Hispanic women see the political arena as a means to enhance the quality of life in their communities and as a way to elicit community participation in the political dialogue pertaining to local environmental and social justice issues. Her data indicate that the majority of female activists see themselves as "third world environmentalists" with a political agenda that includes "the cultural values, beliefs, and life situations of people of color." They frame their political objectives in terms of community empowerment, improved socioeconomic conditions, and sustainable natural use.

PROSPECTS FOR THE FUTURE

In part IV, we focus the reader's attention on the future. We show that coalition building and collaboration among agencies and communities can be successful in addressing health issues. As the physical environment of marginalized communities improves, so too, we believe, will public health. We look forward to proactive and supportive public policies that challenge the systemic causes of social and ecological injustices and help to eliminate health disparities. How will policy makers confront future problems given the options and incentives they face? New questions arise as we look forward to the next generation of reform and emerging issues in the quest for environmental justice. The authors of chapter 9, "In Pursuit of Healthy and Livable Communities," discuss the ways in which both individual and community health are affected when physical environments are unhealthy. They point out that as awareness of environmental health risks have increased, the public has called for more response by the public health sector. Kenny Foscue describes a state-level Brownfields initiative for addressing public health issues at urban sites. He then turns his attention to another public health issue, that of environmental hazards to which people are exposed through indoor air. He discusses the importance of collaboration among agencies and organizations as they work to implement the "Tools for Schools" program in Connecticut. Estelle Bogdonoff explains the principles of coalition building, used by the Southeastern Connecticut Indoor Air Quality Coalition to address indoor air-quality problems in New London County. Kathy Cooper-McDermott explains "Putting on A.I.R.S" (asthma indoor risk strategies), a home intervention program aimed at reducing the burden of asthma among city residents by improving the environments in which they live, play, attend school, and work.

With respect to future prospects, Christopher Foreman in chapter 10, "Three Political Problems for Environmental Justice," writes about political problems confronting policy makers and activists who seek to develop policies that promote environmental justice. From Foreman's perspective, these challenges include issues of environmental risk, remedies for perceived injustices, and the "political sustainability" of the environmental justice movement. He questions whether policy makers can harmonize the "rationalizing and democratizing orientations" that currently compete in the discourse about environmental justice reform. Foreman calls on environmental justice groups to replace their approach of "risk and racism" with a candid discussion of "priorities and tradeoffs." He sees gains from having the environmental justice movement focus on the most severe environmental risks confronting minority communities. In this respect, Foreman argues that policy makers should abandon their approach of making "everything" a priority and concentrate on the long-term interest and needs of minority communities.

In selecting chapters for this book, we chose to take a comprehensive approach to the quest for environmental justice for all communities. Our intention is to offer a book of readings that will facilitate discussion, spark enthusiasm, and motivate readers who have interests in a broad array of topics, such as environmental movements, community participation in decision making, risk analysis, advocacy, and ethics. As a result, we must acknowledge that many topics are included in *Our Backyard* that merit longer treatment than a single chapter can afford. We refer the interested reader to the notes at end of each chapter and to the bibliography for more in-depth study of the topics introduced here.

NOTES

1. R. D. Bullard, "Introduction," in *Confronting Environmental Racism: Voices from the Grassroots,* ed. R. D. Bullard (Boston, Mass.: South End, 1993), 8.

2. C. H. Foreman Jr., *The Promise and Peril of Environmental Justice* (Washington, D.C.: Brookings Institution Press, 1998), 10–11.

3. K. A. Manaster, *Environmental Protection and Justice* (Cincinnati, Ohio: Anderson, 2000), 158.

4. R. D. Bullard, *Dumping in Dixie: Race, Class, and Environmental Quality* (Boulder, Colo.: Westview, 1990), 29–30.

5. General Accounting Office, *Siting of Hazardous Waste Landfills and Their Correlation with Racial and Economic Status of Surrounding Communities* (GAO/RCED-83-168) (Washington, D.C.: General Accounting Office, 1983), 1.

6. United Church of Christ, Commission on Racial Justice, *Toxic Waste and Race in the United States: A National Report on the Racial and Socio-Economic Characteristics of Communities with Hazardous Waste Sites* (New York: United Church of Christ, 1987), xiii.

7. D. L. Anderton, A. B. Anderson, J. M. Oakes, and M. Fraser, "Environmental Equity: The Demographics of Dumping," *Demography* 31 (May 1994): 229.

8. Foreman, *Promise and Peril,* 108.

I

ENVIRONMENTAL HAZARDS IN POOR AND MINORITY COMMUNITIES

1

History and Issues of the Environmental Justice Movement

Bunyan Bryant

Over the years, the terms *environmental racism, environmental equity,* and *environmental justice* have become increasingly used and have grown to mean different things to different people. These concepts, particularly the latter, have gone from obscurity to normality for a considerable number of activists and policymakers within the past decade. The use of such concepts signals a new way in which society as a whole will view the disproportionate health effects of environmental insults upon people based on their biological characteristics or on where they live. Before a discourse on the history of environmental justice can begin, the terms must be defined. Reverend Ben Chavis of the United Church of Christ Commission for Racial Justice first defined *environmental racism* as an extension of racism. *Environmental racism* refers to those institutional rules, regulations, policies, and government or corporate decisions that deliberately target low-income and people-of-color communities for locally undesirable land uses. Environmental racism is the unequal protection against toxic and hazardous waste exposure and the conscious or unconscious systematic exclusion of people of color from environmental decisions affecting their communities.

Environmental equity refers to the equal protection of environmental laws. Abandoned hazardous waste sites in minority areas take 20 percent longer to be placed on the national priority list of the Superfund cleanup program than do those in white areas. The government fines are six times greater for companies violating the Resource Conservation and Recovery Act in white versus black communities. This is unequal protection. Laws should be enforced equally to ensure proper siting of hazardous waste disposal, hazardous waste cleanup, and effective regulation of industrial pollution regardless of the racial and economic composition of the community.

Environmental justice is broader in scope than environmental equity. The term refers to those cultural norms and values, rules, regulations, behaviors, policies, and decisions that support sustainable communities where people can interact with confidence that their environment is safe, nurturing, and productive. Environmental justice is served when people can realize their highest potential without experiencing the "isms" (such as racism). Environmental justice is supported by decent paying and safe jobs, quality schools and recreation, decent housing and adequate health care, democratic decision making and personal empowerment, and communities free of violence, drugs, and poverty. These are communities where both cultural and biological diversity are respected and highly revered and where distributed justice prevails.

In the winter of 1990, a small group of scholar-activists gathered at a retrieval/dissemination conference[1] held at the University of Michigan School of Natural Resources and Environment in Ann Arbor to present papers and discuss the topic of "Race and the Incidence of Environmental Hazards: A Time for Discourse." This conference resulted in two nationally significant events. The first was the publication of a book of readings entitled *Race and the Incidence of Environmental Hazards: A Time for Discourse*, which has been used in environmental and law curricula throughout the country. The second was a series of meetings with Environmental Protection Agency (EPA) administrator William K. Reilly with the goal of providing better protection against environmental harm, particularly for low-income and people-of-color communities. A subgroup of Michigan conferees met with Reilly to express deeply held concerns about the disproportionate impact of environmental hazards on low-income and people-of-color communities. Although community activists had previously confronted EPA regarding its inaction in communities overburdened with toxins, this subgroup of the Michigan Conference[2] challenged the agency to place environmental equity on its radar screen. Administrator Reilly and about twenty of his top administrators listened to the Michigan group presenting its case. Approximately eighteen months later, EPA met this challenge by having the Michigan group return to Washington, D.C., to critique a two-volume report entitled *Environmental Equity: Reducing Risk for All Communities*, which chronicled the environmental equity activities and shortcomings of the agency. Administrator Reilly also created the Environmental Equity Office, which later was renamed the Office of Environmental Justice. These and subsequent meetings with Reilly and his administrators proved to be fruitful, although occasionally heated. The meetings continued with EPA administrator Carol Browner of the Clinton administration, until the National Environmental Justice Advisory Council (NEJAC) was formed.

To understand the history of the environmental justice movement, it must be placed in the context of the civil rights movement and other social and economic events of the 1960s. When Congress passed the 1964 Civil Rights Act and the 1965 Voting Rights Act, many of the white civil rights activists felt their jobs were completed, and they moved on to the peace, environmental, women's, and countercultural movements, all of which have been described loosely as the New Left movement. These individual movements competed for the attention of white activists, as the surge of black nationalism forced whites from leadership positions in the civil rights movement. Black nationalists of the civil rights movement felt that only blacks should direct the course of civil rights organizations and that black leaders would provide models of black pride for people across the generations. In addition to this turmoil, the assassinations and deaths of black leaders shook the movement to its core. Martin Luther King Jr. of the Southern Christian Leadership Conference, Malcolm X of the Organization of Afro American Unity, and Fred Hampton of the Black Panthers were all assassinated in the 1960s. Whitney Young of the National Urban League, A. Phillip Randolph of the Brotherhood of Sleeping Car Porters, and other black leaders died from natural causes. The movement found itself leaderless and adrift like a boat without a rudder.

The environmental movement came of age before the civil rights movement peaked in the 1960s. Rachel Carson's *Silent Spring*,[3] perhaps the greatest influence on the movement, warned of the dangers posed by post–World War II chemicals such as pesticides and herbicides. In addition, Stewart Udall's *The Quiet Crisis*,[4] Paul Ehrlich's *The Population Bomb*,[5] Barry Commoner's *The Closing Circle*,[6] Donella Meadows's *Limits to Growth*,[7] and others played major roles in shaping the environmental movement. This movement rose to prominence at a startling speed. In a short time, the issues of air and water pollution grew from the concern of a small minority of the public to the second-biggest concern of Americans, second only to crime.[8] National environmental organizations took on new life and the new environmental consciousness propelled college students to protest against those forces responsible for waste and environmental degradation. The 1970 Ecology Teach-In took place on college campuses throughout the country and clearly demonstrated that the environmental movement had come of age.

Richard Nixon became a part of this movement when he created the Environmental Protection Agency, and Congress joined in by passing the 1970 Clean Air Act and the 1972 Clean Water Act. Also in 1970, Nixon established the Council on Environmental Quality (CEQ) as a part of the National Environmental Policy Act. CEQ examined the underlying causes of environmental problems in the United States, made recommendations to the president for achieving environmental goals, and in 1971, examined the environmental quality of inner cities. Inadequate housing, high crime rates, poor health and sanitary conditions, inadequate education and recreation, and rampant drug addiction were described in the report.[9] This document was only reporting what was written in the *Black Scholar* the year before.[10] The inner-city environmental concerns raised by CEQ and Nathan Hare, a black sociologist, faded from the national decision-making agenda partially because, in the 1970s, the mainstream environmental organizations were preoccupied with the body politic of Earth Day, preservation, conservation, and environmental aesthetics. While the body politic at the time failed to address the environmental protection of inner cities, a group of scholars managed to keep the spark alive in academia. Several studies found that a relationship existed between race and the disproportionate burden of environmental insults.[11]

In 1976, the University of Michigan School of Natural Resources and Environment played yet another role in advancing the environmental justice movement. Two students from the Environmental Advocacy Program,[12] Phil Perkins and Margaret Allen, worked as staff members for the United Automobile Workers (UAW) Department of Conservation. These students played a major role in organizing a national conference at the Walter and May Reuther UAW Family Education Center held at Black Lake, Michigan. In 1960, this camp cost $30 million dollars to build for training workers to become skilled union organizers.[13] The camp is located in the lake-laced region of northern Michigan. Odessa Komer, UAW vice president and director of the Department of Conservation, opened the conference, entitled "Working for Environmental and Economic Justice and Jobs."[14] This was the first time the term *environmental justice* was used by people in the movement. More than 140 labor, environmental, and community organizations sponsored this five-day conference. Seldom have such divergent groups of people met with more willingness to find common ground. Attendees included Native Americans, Chicanos, blacks, whites, the League of Women Voters, building tradesmen, Appalachian mountaineers, and UAW workers. People came to advance their positions and to listen to one another's points of view. Union officials, ecology activists, and community leaders explored their common interests and attempted to ease the friction between environmental and economic progress. The task was not easy. Leonard Woodcock, president of UAW, and other leaders said that organized labor and environmentalists had a common cause in fighting corporate threats to close down factories or to move them to distant ports as a way of forcing workers and communities to choose between having jobs and ending pollution.[15] The conferees left with a better understanding of one another's jobs and environmental positions. The conference was the precursor to a number of later conferences[16] that helped the environmental justice movement get where it is today.

Two years later, an event in Warren County, North Carolina, shook the black community. Although Love Canal and Times Beach[17] were serious tragedies, the Warren County incident had a more profound impact because it united two different movements. For years, the civil rights movement had been floundering, but the burial of contaminated soil in an African American community provided activists an opportunity to champion civil rights through a different language or frame.[18] Activists began to frame civil rights issues more precisely in terms of health risks related to chemicals and hazardous waste, risk management, soil structures, and landfills. The concrete conditions faced by residents of Warren County provided opportunities for framing environmental and social justice facts to create a different story than had been told in the past. This was the first time that blacks combined the frames of the civil rights and environmental movements to build the contemporary environmental justice movement.

The Warren County incident began in 1978, when the Ward Transformer Company illegally sprayed approximately 31,000 gallons of PCB (polychlorinated biphenyl) fluid on the roadsides of fourteen different North Carolina counties, covering more than 280 miles. When Governor Jim Hunt realized the problem, he had to decide what to do with the contaminated roadside soil. He considered trucking it to Emelle, Alabama, to the largest chemical waste facility in the country. Emelle is located in Sumter County, one of the poorest counties in the nation, where the population is about 75 percent black. Chem Waste Management, the owner of the disposal site, receives waste from forty-eight states and three foreign countries. Over the years, environmentalists have tried to shut down the facility, but the battle has been difficult because Chem Waste Management pays about half of the county's taxes and contributes to black churches and other civic organizations. A number of blacks favor the facility because it provides well-paying jobs. Environmentalists who have tried to close the facility because of potential health risks have been at odds with this black community in desperate need of jobs. In 1978, the plan to ship the waste across country proved too costly, so Governor Hunt decided to bury the soil in Warren County, a predominantly African American community. When the black community found out about the governor's intentions, they quickly organized a protest to stop the trucks from carrying the contaminated soil to the local landfill.[19] Dollie Burwell, a member of the Oak Level United Church of Christ and the Southern Christian Leadership Conference, played a major role in getting people to protest against the disposal of the PCB-contaminated soil in the local dumpsite.[20]

More than five hundred people were arrested in Warren County at the dumpsite, and this event played a major role in starting the contemporary environmental justice movement. The Warren County confrontation drew civil rights activists from all over the country to demonstrate nonviolently against the disposal of the contaminated soil in the predominantly black community and to show solidarity. The demonstration was like a who's who in civil rights,[21] as many prominent activists participated. Although the battle was lost, the outcome of this demonstration clearly indicated that African Americans needed to become involved in electoral politics as a way of protecting their communities against the onslaught of hazardous waste disposal. This event raised the consciousness of black Americans across the country to the potential health effects of hazardous waste disposal sites in their communities.

As a result of this incident, U.S. Representative Walter Fauntroy of Maryland, one of the demonstrators arrested in Warren County, and Representative James Florio of New Jersey commissioned the General Accounting Office (GAO) of the U.S. government to undertake a study of EPA Region IV, the southeastern part of the United States, which includes Warren County. The goal was to see if Warren County was unique with regard to hazardous waste disposal sites. The 1983 study showed that three of the four largest landfills in EPA's Region IV were in predominantly black communities. Even when income was controlled, race was still a greater explanatory variable in the location of commercial hazardous waste sites in Region IV African American communities. Little was made of this study, but in 1987 the United Church of Christ's Commission for Racial Justice,[22] under the leadership of Reverend Ben Chavis, contracted to have a national study conducted to see if a pattern existed between where waste sites were located and where people of color lived. *Toxic Waste and Race in the United States* was the first national study to state that race proved to be the most significant factor among variables tested in the location of hazardous waste facilities, even when controlled for income, regional differences, and urbanization. The results of this study[23] were similar to the results of the GAO study. To date, this national study has been the most cited study in professional and lay journals and at conferences, and has had a tremendous impact upon communities across the nation. The GAO and the *Toxic Waste and Race in the United States* studies were precursors to the 1990 Michigan Conference on "Race and the Incidence of Environmental Hazards: A Time for Discourse."

PEOPLE OF COLOR AND THE ENVIRONMENT

For a long time environmentalists criticized people of color for their conspicuous absence from the environmental movement. People of color responded to this criticism by saying they seldom had time for the environmental movement as defined by whites. People of color claimed they had more pressing issues of survival to contend with in their neighborhoods. In reality, people of color have always been concerned about environmental issues. They have expressed concerns about toxic waste sites and ambient air and water quality. They championed decent recreation facilities, well-paying jobs, decent housing, safe streets, and quality schools. The 1990 Detroit Area Study found that a higher percentage of blacks than whites rated quality-of-life issues as serious and that blacks seemed to be more aware of and concerned about environmental problems around them.[24] A 1973–1988 study at the University of Chicago reported that, in every year except one (1985), a larger percentage of blacks than whites felt that the nation spent too few dollars to protect the environment.[25] Although people of color may not have joined white environmental organizations, they had an interest in environmental issues that they have also viewed as survival issues.

The environmental track record of leaders of color is impressive.[26] Cesar Chavez fought to improve the environmental conditions for farm workers, although he did not frame the exposure to harmful pesticides as an environmental justice issue. Nathan Hare castigated the white environmental movement in 1970 for failing to consider the urban environment as a cause for concern. In 1978, Vernon Jordan of the Urban League stated that appropriate technology was good for the black community because it provided environmentally benign jobs. Appropriate technology is technology that is used to meet the world's needs in a sustained, energy efficient, and environmentally benign way. This includes affordable and energy-efficient systems, bioregional development, and community and shelter design. It includes technology that uses alternative forms of energy, namely solar, geothermal, and recycling. It is the opposite of capital intensive and energy inefficient technologies that substitutes energy for labor. In 1984 and 1988, presidential candidate Jesse Jackson had a stronger environmental platform than those of his Democratic rivals. For many years, members of the Congressional Black Caucus have had better environmental voting records than their white counterparts, according to data provided by the League of Conservation Voters.[27] Representative John Conyers of Detroit authored bills to curtail transportation of hazardous waste to developing countries, or to at least make it safe, and he authored a bill to make EPA a cabinet-level position. These are only a few examples of leaders of color who have worked diligently to protect the environment. Also, there are countless activists across the nation working to improve the quality of life where they live. Although professionals and universities have played a role, this movement could not be what it is today without the work of these unsung heroes and heroines who risk their lives so that others can live in safe, nurturing, and productive communities. Grassroots activists have been instrumental in shaping the environmental justice movement at local, national, and international levels.

CONTEXTUAL HISTORY OF THE 1990S

A contextual view of the 1990s is helpful in understanding how state governments and industry have viewed the environmental justice movement. Since the early 1970s, industry has fought against environmental regulations, claiming they stifle creativity.[28] Arguments against regulations asserted that adhering to environmental regulations cut into profit margins and thus placed U.S. industry at an unfair market advantage with unregulated foreign competitors.

Workers were pressured to side with industry against environmentalists and regulations or risk losing their jobs, a practice known as environmental blackmail. This issue was addressed at the 1976 conference on "Working for Environment and Economic Justice and Jobs." For years, industry has attacked environmentalists and regulations, even though the overall economy has continued to improve. Japan and Germany have the most stringent environmental regulations in the world, yet these countries have experienced robust economies, particularly in the 1980s. In the 1990s, the United States experienced the strongest economy in history, and yet industry continues to complain about environmental regulations. Why the resistance? Perhaps the problem is that industry is afraid of change even when that change is in its long-term best interest.

Six concerns in the 1990s piqued the interest of industry and the state in the environmental justice movement. The first concern was the equal protection of the law as a constitutional guarantee. Although equal protection of the law is a doctrine of our constitution, it is seldom applied to communities overburdened by environmental hazards. The issue of equal protection against environmental harm was seldom raised until the environmental justice movement was formed. Industry was afraid of adhering to equal protection because this meant spending money on pollution abatement and control technology, and on finding ways to make plants safe for workers and people in the surrounding communities. The second concern was the establishment of the Office of Environmental Justice, which was given a mandate to (1) serve as a focal point to ensure that people-of-color or low-income populations receive protection under environmental laws, (2) provide oversight regarding these concerns relating to all parts of environmental protection, and (3) work in a coordinated effort with other agencies to ensure environmental protection. The third concern was the creation, under Clinton's EPA administrator, Carol Browner, of NEJAC, which consisted of people from different sectors of society who were to advise Browner on environmental justice policies. The fourth concern was the issuing of the 1994 Environmental Justice Executive Order 12898, requiring that each federal agency make achieving environmental justice a part of its mission by identifying and addressing disproportionately adverse human health or environmental impacts of its programs, policies, and activities on people-of-color populations and low-income communities in the United States. The fifth concern was cumulative impact, requiring that an industry do an assessment of the total impact of industries already located in an area before it is permitted to site there. Such as assessment can help determine if additional impacts will be within a permissible threshold.[29] The sixth concern was the creation of the 1998 Title VI guidance,[30] which has been used as a tool to protect people-of-color communities from environmental harm. In Michigan, more so than in other states, the environmental blackmail rhetoric came back in full force when hearings were held on Title VI. The governor, the director of environmental quality, and the former mayor of Detroit came out against Title VI, claiming it was a disincentive to investment and would cost thousands of jobs, even though research showed the contrary. These six concerns are viewed by industry as disincentives to investments that cut into their profit margins rather than provide economic opportunities.

KEY ENVIRONMENTAL JUSTICE CONFERENCES OF THE 1990S

One year after the 1990 Michigan Conference, Charles Lee of the United Church of Christ Commission for Racial Justice organized the "First National People of Color Environmental Leadership Summit," held at the Washington Court Hotel in Washington, D.C. More than five hundred people from all over the country, mostly people of color, attended this summit. Leaders of two of the most prominent environmental organizations, Michael Fischer, executive director of the Sierra Club, and John Adams of the Natural Resources Defense Council, addressed the audience. Both men expressed sympathy for the environmental problems of minority communities, acknowledged the conspicuous silence of their organizations in addressing people-of-color environmental issues, and appealed for racial unity, but the appeal for unity was not enough. In both small- and large-group discussions, people of color gave emotionally charged testimonies about the poisons affecting their communities. The document *17 Principles of Environmental Justice* came out of this historic summit and was to be used as an organizing tool.[31]

A second event of note was the 1994 "Health Research and Needs to Ensure Environmental Justice Conference," sponsored by the EPA, the Agency for Toxic Substances and Disease Registry (ATSDR), the National Institute of Environmental Health Sciences, the Department of Energy, and the Centers for Disease Control and Prevention. More than one thousand people, at least one-third from grassroots organizations, spent three days at the conference outside of Washington, D.C. At times, the conference was confrontational. People shouted that they didn't need more studies because they knew the problems of their communities. They just wanted the financial resources to do something about them. Recommendations from this conference included the funding of meaningful community-based research projects, promoting disease and pollution prevention strategies, supporting new ways of interagency coordination in order to be more helpful to communities, promoting community outreach, education, and communication, and using effective legislative and legal remedies for environmental protection. In summary, this conference supported community-based research, giving it legitimacy as a recipient of government funding.

KEY ENVIRONMENTAL JUSTICE
RESEARCH FINDINGS AND DEBATES[32]

Three key studies have helped shape the environmental justice debate. These studies are important because of their policy implications, which can significantly impact communities of color overburdened with toxins for years to come. Vicki Been[33] states that locally undesirable land uses (LULUs) are disproportionately located in communities of color and low-income communities. Been claims that her research does not establish that host communities were disproportionately minority at the time the sites were selected. She raises the question of whether the disproportionate burdens of LULUs were due to siting decisions or whether they were due to discrimination in the housing market. For example, if people of color or people with low incomes disproportionately populated the neighborhood at the time of a LULU siting, then a reasonable inference would be to change the siting policy. However, if people moved into an area after the LULUs were built and the area then became increasingly poor or increasingly populated by people of color, then the solution would be more complex. The distribution of LULUs appears more like the age-old problems of housing discrimination and poverty. Additionally, Been speculates that communities surrounding LULUs may become more inhabited by people of color and low-income people due to poverty, job location, and lack of transportation and public services. As the percentages of minorities and low-income people increase in these communities, the likelihood of experiencing discrimination in zoning, environmental protection, municipal service provisions, and bank-lending practices also increases. These forces contribute to additional decline in their communities.

Two other studies have contributed to the national debate. One study was conducted by the Commission for Racial Justice (CRJ); the other was conducted by Douglas Anderton of the University of Massachusetts. Anderton's study challenged the often-cited CRJ study. These two studies are important because both are national in scope and examine the distribution of commercial hazardous waste facilities, yet their outcomes are significantly different. CRJ found that, among the variables tested, race proved to be the most significant in relation to the location of commercial hazardous waste facilities. Although socioeconomic status appeared to play an important role in the location of these facilities, race still proved to be more significant, even when the study was controlled for urbanization and regional differences. However, through the use of census tracts, Anderton's U. Mass. study found that treatment storage and disposal facilities (TSDFs) were no more likely to be located in tracts with high percentages of blacks than they were to be located in other tracts. However, these facilities were located in areas that had slightly higher percentages of Hispanics than other tracts. This study reported that TSDFs were more likely to be sited in industrial tracts, and in general those tracts do not have a greater number of minority residents.

Why did these two studies examining the same problem have such different conclusions? A closer examination reveals that the methodology used in these two studies differed significantly, which could explain the diverse results. The CRJ study used zip codes as the unit of analysis, while the U. Mass. study used census tracts. The difference in unit of analysis alone is enough to yield different outcomes. However, there were additional differences. The CRJ included all people of color in its study, but the U. Mass. study included only blacks and Hispanics. The U. Mass. study also considered percentages of males employed in the civilian labor force and percentages of males employed in the manufacturing and industry fields. The CRJ did not consider such percentages.[34]

These studies are important in considering the national debate and policy making in environmental justice and the backlash taking place today. Been's study becomes a chicken-and-egg argument of which came first, the people or the sites? This argument leads to a paralysis of analysis. Indeed, there is racial discrimination in siting and in the housing market; the black community has known this for years. Whether the hazardous waste sites were there before or after the people does not matter. What matters is that people are overburdened with toxins and often live in life-threatening conditions. A policy protecting all people regardless of race or income must be formulated and acted upon. The Anderton study often provides fuel for the environmental justice backlash. This is one of the few studies that attempts to refute the outcomes of a number of studies that state that the distribution of environmental burdens and benefits is associated with race.

In 1992, Paul Mohai and Bunyan Bryant showed that an overwhelming majority of sixteen case studies they reviewed found that the distribution of the environmental burden was inequitable by income regardless of the environmental hazard and the scope of the study. This was true in all cases except one.[35] Where it was possible to weigh the relative importance of each, race in most cases was found to be more strongly related to the incidence of pollution, even when race and income were combined in the analysis of pollution. In 1994, Benjamin Goldman examined sixty-four studies and found that racial disparities were more frequent than income disparities (87 percent of the tests showed racial disparity compared with 74 percent that showed income disparity). Race proved to be the more significant factor in 75 percent of the tests when race and income were combined, as in the Mohai and Bryant study. A national study using primary data would be useful because most studies to date use secondary data sources. The debate continues, even though most studies clearly show race as a greater explanatory variable than income in the association of environmental burdens.

THE LEGISLATIVE ROUTE FOR ENVIRONMENTAL JUSTICE

Several legislative attempts at improving environmental justice took place in the early to mid-1990s. In 1992, the House Committee on Health and Environment held hearings on the impact of lead poisoning and its effect on low-income and people-of-color communities. This hearing was used effectively by environmental justice advocates to present their cases to Congress and to set the groundwork for future legislation. At committee hearings on Capitol Hill, Reverend Ben Chavis of the United Church of Christ Commission for Racial Justice supported the Environmental Justice Act of 1992. He advocated that the proposed legislation limit future siting of pollution facilities in one hundred high-impact areas across the country and that government money should be made available to enhance public participation in these areas. In November 1993, Chavis, as the director of the National Association for the Advancement of Colored People (NAACP), testified before the House Subcommittee on Transportation and Hazardous Materials on the need for new, stand-alone environmental legislation to address the issue rather than going through existing legislation. At this hearing, Robert Bullard suggested a four-prong strategy for addressing environmental justice that included (1) enforcement of existing laws and regulations, (2) an environmental justice executive order, (3) new federal legislative initiatives, and (4) state and local legislation. At these hearings, Deeohn Ferris of the Lawyer's Committee for Civil Rights Under the Law supported three House bills that dealt with environmental justice. Ferris also called for the federal government to conduct a comprehensive evaluation of environmental health impact areas and for the government to craft an interagency group to address environmental justice issues internally.[36]

At the June 28, 1994, hearings before the Senate Subcommittee on Public Works on the authorization of Superfund,[37] Chavis outlined a program to aid and strengthen Superfund in order for it to be more responsive to environmental justice issues. Points raised by Chavis in his testimony were developed through a cooperative effort between the NAACP and the Alliance for a Superfund Action Partnership (ASAP), which includes members from the corporate sector, local government, and grassroots organizations. Florence Robinson's testimony at these hearings was in favor of retaining the retroactive liability clause or the polluter pay bill. Her remarks differed from Chavis's remarks made on behalf of ASAP.

In 1990, the Congressional Black Caucus provided a forum for environmental justice during one of its legislative weekends. The Congressional Black Caucus forum is an annual event that draws blacks from multiple professions and all walks of life to participate in workshops addressing issues unique to the black experience. These sessions not only educate black congressmen, but also motivate them to support and submit environmental justice legislation. In 1991, Representative John Lewis introduced his Environmental Equity Act

(H.R. 5326) as a stand-alone piece of legislation. Environmental justice is both a civil rights issue and an environmental protection issue, a fact Representative Lewis kept in focus. To address multiple exposures to pollutants, Representative Candis Collins of Illinois introduced H.R. 1924,[38] but the political climate was not supportive of such legislation. Perhaps some time in the future, this legislation or a modified version will make its way back into the limelight.

ENVIRONMENTAL JUSTICE STRUGGLES

Throughout the United States, activist groups are involved in struggles to protect their communities from environmental harm.[39] One such struggle culminated on March 15, 1990, when members of Louisiana's Gulf Coast Tenant Leadership Development Project and the Southwest Organizing Project sent a letter to the Big Ten environmental organizations, criticizing them for having white, homogeneous staffs and for accepting money from corporations that were poisoning and killing people-of-color communities, the very communities that environmental justice activists were trying to protect.[40] However, most of the environmental justice struggles in the country do not focus on the Big Ten national environmental organizations. Most struggles in this country take place at the local level. Local activists are involved in a variety of struggles involving incinerators, sewage treatment plants, oil refineries, uranium mines, hazardous waste depository sites, concentrated animal feed organizations (i.e., hog farms, because of pungent odors and soil and water contamination), and landfills. Community activists have focused on truck and marine transfer stations where garbage is collected for shipment, nuclear power plants and the disposal of nuclear waste, government waste, water- and land-rights issues, and agricultural runoff. Low-income communities and communities of color across the country have fought against a government that has failed to make polluting industries accountable and against companies that poison ambient air, water, and land. They have fought against companies that have wanted to use Native American reservation land for the disposal of toxic and hazardous waste. The story is the same no matter where one goes. People-of-color and low-income communities feel they are recipients of environmental burdens, but not the benefits. Some of the most notorious battles for environmental health[41] have been with the oil refineries and petrochemical plants that line the eighty-mile strip along the lower reaches of the Mississippi River between New Orleans and Baton Rouge. This strip was named Cancer Alley by the predominantly African American community living there.

Thousands of activists across the country struggling to protect their communities against poisons have experienced degrees of success.[42] Activists have been most successful when they use multiple strategies such as mobilizing the community for direct action, nonviolent demonstrations combined with media usage, scientific evidence, and legal tactics. However, an environmental solution for one community often becomes a problem for another community, as polluting industries move from community to community. For example, considerable opposition was expressed against Shintech, a company that planned to build a plant in Convent, Louisiana. The company reacted by moving into another community. This also happened with Select Steel in Flint, Michigan, where opposition forces were so great that the company moved to another area. To avoid this, we must continue the struggle to find outcomes similar to that of the following case study.

THE HENRY FORD MEDICAL WASTE INCINERATOR STRUGGLE

Of all the environmental struggles in the past twenty years, few if any have had the significance of a little-known struggle that took place against Detroit's Henry Ford Medical Waste Incinerator. On February 4, 2000, Henry Ford Health System (HFHS) announced the shutdown of its medical waste incinerator at the Detroit Henry Ford Hospital. The hospital, located in a mixed business and residential community consisting primarily of African Americans, is one of the largest in Michigan. The incinerator burned 600 million pounds of waste annually. The rate of children hospitalized for asthma is at least three times higher in the inner-city zip codes surrounding Henry Ford Hospital than it is in Wayne County or central Detroit. The asthma rate among African American children living near the hospital seems to be increasing, as reported in the Wayne State University study.[43] The decision to shut down the incinerator came as a result of two years of protests against its operation and the health hazards it posed to the community. Although the coalition of community and environmental justice activists congratulated the Henry Ford Hospital administration for making the decision to close the incinerator, they also pushed for a definite and expedient time line. During the struggle, the coalition raised issues about the inconsistency of HFHS conducting asthma research and providing treatment for the disease while supporting smokestack emissions from an incinerator that perhaps contributed to rising asthma rates. Henry Ford Medical Center could no longer live with the contradictions of being healer and polluter. This case is also significant because Henry Ford Hospital did not choose to build a replacement incinerator in another community.

The mobilization of community and environmental justice activist groups was a key factor in the decision to close HFHS's Detroit incinerator. Eleven community, environmental, health, and citizens groups were involved in bringing about the shutdown of this incinerator. Demonstrations in front of the hospital educated and broadened the arena of conflict by bringing more people to the side of the protesters. Demonstrators used science to show the harmful effects of dioxin and other chemicals emitted from the incinerator smokestack and thus were able to stake out the moral high ground. Science was also used to present HFHS with alternative technologies such as steam sterilization and autoclaving, which would be safer and perhaps cheaper in the long run. After HFHS administrators made the decision to close the incinerator, they consulted with the community coalition to gain names of experts to help with the transition process and to provide support of a proposal to fund the transition. This became a collaborative process in which the administrators treated the coalition members with respect and integrally involved them in the problem-solving process. While this activism had a positive result, focus on the front end of the production cycle is still necessary to prevent or severely limit toxins from occurring in the first place and, ultimately, to be more effective in protecting all communities regardless of race, income, national origin, or gender.[44]

THE BACKLASH

Not all was well with the environmental justice movement in the late 1990s. Industry, which has a history of resisting environmental regulations, claiming that they place undue burden upon business and thus cut into profit margins, took up the charge again, this time against the environmental justice movement. To protect low-income and people-of-color communities against environmental harm, environmental justice (EJ) activists not only pressured EPA to apply the equal protection of the law to ameliorate environmental conditions but also, in the late 1990s, were able to get EPA to explore and apply Title VI of the 1964 Civil Rights Act to protect people-of-color communities against environmental harm and encourage industry to use cumulative impact in its siting decisions. Lastly, EJ activists were able to get President Clinton to issue the 1994 Environmental Justice Executive Order 12898, which required all federal agencies to take stock of its EJ activities and to craft and implement a plan of action. What industry did in the environmental movement it did in the EJ movement in that it went on the offense, claiming that environmental rules and regulations were a disincentive to economic investment. The environmental justice leadership felt attacked for supporting rules and regulations that protected low-income and people-of-color communities.

On December 12, 1999, the National Emergency Gathering of Black Community Advocates for Environmental and Economic Justice met in New Orleans to address reactionary forces trying to turn back the clock of environmental justice. Scholars and activists had been attacked in the media and in the courts for their work to ameliorate detestable environmental conditions in low-income communities and communities of color. The opposition had sued environmental justice advocates in court. This is known as strategic lawsuits against public participation (SLAPP), a tactic that intimidates and disquiets activists. In most cases, the opposition loses in court, but by that time activists have spent considerable amounts of time and money defending themselves. This indeed could have a disquieting effect on future activities.

The National Black Chamber of Commerce, the Chemical Manufacturers Association, the National Association of Manufacturers, the Washington Legal Foundation, some Republicans and Democrats, and the media have launched a campaign to discredit the environmental justice movement and the environmental justice policy framework. The gathering of the Interim National Black Environmental and Economic Justice Coordinating Committee in New Orleans was established to defeat this reactionary effort and to develop a proactive strategy for advancing the cause of environmental justice in the black community. One of many activities of the Coordinating Committee was planning the May 2–7, 2001, meeting in Detroit to assess, at its halfway mark, the action campaign developed at the New Orleans gathering.

Gathering in Detroit symbolizes the many reactionary environmental justice struggles that have taken place in Michigan. Governor John Engler, former Detroit Mayor Dennis Archer, and Director of the Michigan Department of Environmental Quality Russ Harding have spoken out against environmental regulations and the Title VI Guidance. The *Detroit News* has attacked environmental justice scholars and activists across the country. Michigan has a history of environmental justice struggles dating back to the Detroit incinerator, the Flint wood-burning incinerator, and the attempt to site the Select Steel Industry in Flint. In many respects, Governor Engler and the State of Michigan epitomize the environmental justice backlash.

Even though the movement has been attacked, it continues to grow in capacity. Today the movement has taken on international significance, as EJ activists champion the cause of climate justice. To raise the consciousness of people, environmental activists have used the United Nations forum or conferences paralleling those of the UN to get out its messages. In November 2000, thousands of activists traveled to the Hague to participate in various briefings and alternative summits regarding the effects of global climate change. These alternative briefings and summits were held in addition to the Conference of Parties (COP 6) of the UN-sanctioned climate convention. Because many believe that climate change may be the greatest environmental injustice of all, activists were there to raise critical climate issues. If or when global climate change takes effect, the fear is that a mass exodus of people from flooded and dry areas in search of food, housing, and employment will take place. The poor will have to pay a higher percentage of their income for the basics in life. The poor will also pay a higher percentage of their income on health care as tropical diseases move farther north. If such migration across geopolitical boundaries occurs, it will cause regional conflicts and world disequilibrium. Because poverty and environmental degradation have increased since the 1992 Summit held in Rio de Janeiro, the United Nations General Assembly authorized the World Summit on Sustainable Development to address many of the world's most pressing problems. Unfortunately the world's Summit on Sustainable Development, held in Johannesburg in 2002 did not move us much past Agenda 21 (a global plan for sustainable development adopted at the 1992 Summit in Rio). Although there was considerable potential to do a lot of good in the world, the needs of the many were compromised to accommodate the demands of a powerful few. Given the alarming tilt towards the self-interest of the more affluent countries, will the concerns of the poor ever be addressed in global negotiations? Only time will tell.

Environmental justice activists must prepare to meet the challenges of reactionary forces that will continue well into the twenty-first century. Although colleges and universities must play a critical role in helping people to protect their communities against poisons, they must not upstage the environmental justice movement. Therefore, we must encourage and train new professionals to humble themselves and to work with communities on a more equal basis. We must do it now. Tomorrow may be too late.

NOTES

1. This conference is known as the Michigan Conference and was organized by Professors Bunyan Bryant and Paul Mohai of the University of Michigan School of Natural Resources and Environment.

2. Bunyan Bryant, Robert Bullard, David Hahn-Baker, Ben Chavis, Charles Lee, Michel Gelobter, Paul Mohai, and Beverly Wright met with Reilly and his administrators (Chavis and Hahn-Baker were not part of the Michigan Conference, but joined the group in Washington). The group, known as the Michigan Group, met with staff member Dale Curtis of the Council of Environmental Quality, who was unfamiliar with environmental equity. The group not only educated him about the issue, but also expressed their displeasure about the conspicuous absence of people of color on his staff. The group later met with Congressman John Lewis from Georgia and a staff member of Representative Ron Dellums. The congressman and the staff member understood the issues brought to them and they were enthusiastic about the mission of the Michigan Group. Congressman Lewis promised to put the political power of the Congressional Black Caucus behind the efforts of the group in order to get the EPA to take environmental equity seriously.

3. Rachel Carson, *Silent Spring* (Boston: Houghton Mifflin, 1962).

4. Stewart L. Udall, *The Quiet Crisis* (New York: Holt, Rinehart and Winston, 1963).

5. Paul Ehrlich, *The Population Bomb* (New York: Ballantine, 1968).

6. Barry Commoner, *The Closing Circle: Nature, Man, and Technology* (New York: Knopf, 1971).

7. Donella Meadows, *The Limits to Growth: A Report to the Club of Rome Project on the Predicament of Mankind* (New York: Universe, 1972).

8. Craig Humphrey and Frederick H. Buttel, *Environment, Energy, and Society* (Belmont, Calif.: Wadsworth, 1982).

9. James P. Lester, David W. Allen, and Kelly M. Hill, *Environmental Injustice in the United States: Myths and Realities* (Boulder, Colo.: Westview, 2001).

10. Nathan Hare, "Black Ecology," *Black Scholar* 1 (April 1970): 2–8.

11. Peter Asch and Joseph J. Seneca, "Some Evidence on the Distribution of Air Quality," *Land Economics* 54, no. 3 (1978): 278–97; Brian Berry et al., *The Social Burdens of Environmental Pollution: A Comparative Metropolitan Data Source* (Cambridge, Mass.: Ballinger, 1977); William R. Burch, "The Peregrine Falcon and the Urban Poor: Some Sociological Interrelations," in *Human Ecology, An Environmental Approach*, ed. Peter Richerson and James McEvoy, 308–316 (Belmont, Calif.: Duxbury, 1976); Clarence J. Davies and Barbara S. Davies, *The Politics of Pollution*, 2d ed. (Indianapolis: Pegasus, 1975): William J. Kruvant, "People, Energy and Pollution," in *The American Energy Consumer*, ed. Dorothy Newman and Dawn Day (Cambridge, Mass.: Ballinger, 1975), 125–67.

12. Bunyan Bryant and James Crowfoot started the Environmental Advocacy Program in the School in 1972 as a student-initiated program with a mission of making a connection between the environment and people. The Environmental Advocacy brochure stated that a high priority was placed on understanding the environmental problems of groups least served by the present sociopolitical system and that an emphasis was put on how individuals, groups, corporations, and governments can be more responsive to social and environmental conditions. For almost thirty years, the School of Natural Resources and Environment has addressed equity and justice issues under the heading of Environmental Advocacy (now called Environmental Justice).

13. Although the school played a role in the success of this event, the conference was organized and driven by people of color, union members, and other groups. This historical event will hopefully stimulate other colleges and universities to get involved in similar endeavors. Clark Atlanta in Atlanta, the University of Michigan in Ann Arbor, Texas Southern in Houston, Florida A. & M. in Tallahassee, and Xavier in New Orleans have been among the most visible universities to date working in the environmental justice field.

14. *Working for Environmental and Economic Justice and Jobs Labor* (Indianapolis: Labor News, June, 1976).

15. Frances D'Hondt, "Jobs vs. Environment: Must There Be a Conflict?" *Solidarity* 19, no. 5 (1976): 6.

16. The 1976 conference held at Black Lake, Michigan, and the earlier the 1970 Earth Day Conference made important contributions to the environmental justice movement. Also, Dana Alston's *We Speak for Ourselves: Social Justice, Race, and Environment* (Washington, D.C.: Panos Institute, 1990) chronicles the history of environmental justice–related conferences.

17. In 1942, Hooker Chemical and Plastics (now Occidental Chemical Corporation) used a landfill in New York state for the disposal of more than twenty-one tons of chemical wastes. Dumping ceased in 1952 and, in 1953, the landfill was covered and deeded to the Niagara Falls Board of Education. In subsequent years, the area near the landfill was extensively developed and included an elementary school and homes. Dioxin and other chemicals migrated into Love Canal's existing sewers, which had outfalls into nearby creeks. Residents were outraged when they found that

illnesses in their community could be related to the chemicals seeping from the land-fill. President Jimmy Carter issued an environmental emergency order for Love Canal, and more than 950 families were evacuated from the area (see www.epa.gov/region02/superfnd/siste_sum/0201290chtm). Chemicals, hazardous waste, and waste oil were mixed and used to spray the streets of Times Beach, Missouri. On February 23, 1983, the EPA announced plans to buy out the entire city, which had a popula-tion of 2,061. At that time, the residents believed they already had enough dioxin in their bodies to cause serious health problems. This led to "Where the Mess at Times Beach Really Came From" (see www.geocities.com/Athens/Styx/3222/index-tb.html). This was an environmental justice issue, but was not framed as such. The Warren County incident, by contrast, was given an environmental justice frame.

18. Framing is the process of deciding which of many daily occurrences will be awarded significance as an important event. In this sense, framing selects and orga-nizes pieces of information into stories that make sense to the organizer and target au-diences. One frames an issue by packaging certain facts and events around it. For more information on framing, see "What Is Framing? Why Is It Important?" *Community Change,* special issue (Spring 1998): 15–17; S. M. Capek, "Environmental Justice Frame: A Conceptual Discussion and an Application," *Social Problems* 40, no. 1 (February 1990); J. Dorsey, *Community-Based Activism within an Environmental Justice Frame: The Siting of a Waste to Energy Facility in Flint-Genesee County* (Ph.D. diss., School of Natural Resources and Environment, University of Michigan, Ann Arbor, 1999).

19. Jenny LaBalme, *A Road to Walk: A Struggle for Environmental Justice* (Durham, N.C.: Regulator, 1987).

20. The black church has been a major organizing institution in its community. The black church is unique in that it emerged from slavery as a stable and coherent institu-tion. Black schools, banks, insurance companies, musical talent, and political action were nurtured by the culture of the church. Although there are many competing insti-tutions of the twenty-first century, the black church still remains the communal back-bone. Sojourner Truth, Harriet Tubman, Mary McLeod Bethune, and licensed preach-ers such as Frederick Douglass, Booker T. Washington, Adam Clayton Powell, Martin Luther King Jr., Jesse Jackson, Ben Chavis, and others came from the black church.

21. Demonstrating or supporting the struggle at Warren County were such nota-bles as Reverend Ben Chavis, Ken Ferruccio, Leon White, Dollie Burwell, and Joseph Lowery, to name a few.

22. The Commission for Racial Justice was formed in 1963 in response to the as-sassination of Medgar Evers, the Birmingham church bombings, and the tensions that gripped the nation. As a Protestant church-based organization, the commission fo-cused on such issues as voter registration, access to quality schools for students of color, racial violence, and other civil rights–related events.

23. Over the years, scholars such as Bunyan Bryant and Paul Mohai (*Race and the Incidence of Environmental Hazards: A Time for Discourse,* 1992), Robert D. Bullard (*Confronting Environmental Racism: Voices from the Grassroots,* 1993) have found that race was a significant factor in the location of these facilities. Bryant and Mohai said race was statistically a greater explanatory variable than income. Commission for Racial Justice (CRJ), *Toxic Waste and Race in the U.S.: A National Report on the Racial and Socioeconomic Characteristics of Communities with Hazardous Waste Sites* (New York: United Church of Christ). Benjamin A. Goldman and Laura Fitton, *Toxic Waste*

and Race Revisited: An Update of the 1987 Report on the Racial and Socioeconomic Characteristics of Communities with Hazardous Waste Sites (Washington, D.C.: Co-sponsored by the Center for Policy Alternatives, the National Association for the Advancement of Colored People, and the United Church of Christ, 1994).

24. Paul Mohai and Bunyan Bryant, "Environmental Racism: Reviewing the Evidence," in *Race and the Incidence of Environmental Hazards: A Time for Discourse,* ed. Bunyan Bryant and Paul Mohai, 163–76 (Boulder, Colo.: Westview, 1992).

25. Paul Mohai, "Black Environmentalism," *Social Science Quarterly* 71, no. 4 (December 1990): 744–65.

26. Today Latina/os, Native Americans, African Americans, and Asian/Pacific Islander Americans carry on the struggle for environmental justice in a variety of networks and organizations, including the Asian Pacific Environmental Network, the Indigenous Environmental Network, the Southern Organizing Committee, the National Black Environmental Justice Network, the Northeast Environmental Justice Network, the Southwest Network for Environmental and Economic Justice, and the Farmworker Network for Economic and Environmental Justice. The purpose of these networks, each consisting of numerous organizations, is to influence local, state, and national environmental and economic policy.

27. Henry Davis, "The Environmental Voting Record of the Congressional Black Caucus," in *Race and the Incidence of Environmental Hazards,* ed. Bryant and Mohai, 55–63.

28. In 1990, President Bush created the Council on Competitiveness, headed by Vice President Dan Quayle, to cut all "unnecessary" government programs (programs that slowed economic growth, including the agencies that enforced the Clean Air Act). See http//usepages.umsbc.edu/~cghrm1/pres_site/presidents/gb.html. Also coming out against environmental regulations was the CONSAD Corporation, an economic and public analysis consulting firm that serves a variety of clients, including the Business Roundtable, the Clean Air Working Group, and many other corporations. CONSAD has supported industry interests, reporting that the Clean Air Act Amendments of 1990 would negatively impact the economy due to job loss.

29. Cumulative impact has been a big issue for states and industry because of the time it takes to conduct a cumulative impact analysis and because of the potential disincentive to investment. For example, if an assessment is under threshold, then location is not a problem. If the impact is too high, however, then the industry cannot locate unless industries already located there are persuaded to reduce their pollution levels.

30. Title VI of the 1964 Civil Rights Act was geared to prevent racial discrimination in federally assisted programs by withdrawing federal funds from companies practicing such discrimination. Title VI guidance is now being used (as guidance—not rule of law) to help avoid siting discrimination or to ameliorate those value-neutral policies resulting in discriminatory effects, particularly if a state or corporate entity is the recipient of government funds. The shortcoming of Title VI is that it only applies when people of color are living disproportionately near toxic waste or polluting facilities. Title VI does not apply to whites who may also be impacted disproportionately by polluting facilities.

31. The preamble to *17 Principles of Environmental Justice* is as follows: "to begin to build a national movement of all people of color to fight the destruction and taking of our lands and communities; to hereby reestablish our spiritual interdependence to

the sacredness of our Mother Earth; to respect and celebrate each of our cultures, languages, and beliefs about the natural world and our roles in healing ourselves; to ensure environmental justice; to promote economic alternatives, which would contribute to the development of environmentally safe livelihoods; and to secure our political, economic, and cultural liberation that has been denied for over 500 years of colonization and oppression resulting in the poisoning of our communities and land and genocide of our people, do affirm and adopt these principles of Environmental Justice. . . ." The principles have been widely used throughout the country.

32. The writings of Bunyan Bryant, Paul Mohai, Dorceta Taylor, Robert Bullard, Beverly Wright, Michel Gelobter, Charles Lee, Luke Cole, Carl Anthony, Bob Kuehn, and others have contributed significantly to this debate.

33. Vicki Been, "Locally Undesirable Land Uses in Minority Neighborhoods: Disproportionate Siting or Marketing Dynamics?" *Yale Law Journal* 103, no. 6 (April 1994).

34. For more information on the methodological differences between the two studies, see Paul Mohai, "The Demographics of Dumping Revisited: Examining the Impact of Alternate Methodologies in Environmental Justice Research," *Virginia Environmental Law Journal* 14 (1995): 615–53.

35. At the time, this was one of the few quantitative studies that showed that race had an effect greater than income.

36. Lester et al., *Environmental.*

37. The Comprehensive Environmental Response Compensation and Liability Act of 1980 is popularly known as the "Superfund Act."

38. Lester et al., *Environmental.*

39. The number of Asian and Pacific Islanders in the United States has grown fourfold in the past two decades, from 1.5 million in 1970 to 7.3 million in 1990, yet this group accounts for only 3 percent of the population. Environmentally, the Asian Pacific community faces many of the hazards that other communities of color do. Nearly half of all Asian Americans, Pacific Islanders, and Native Americans live in communities with uncontrolled or illegal waste sites. Forty percent of the Asian and Pacific Islander community in the United States lives in California, where they make up 10 percent of the state's total population. In the San Francisco Bay Area, 30 percent of the population is Asian and Pacific, according to a recent study (Environmental Justice Issues of Asians and Pacific Islanders in the United States, www.igc.org/envjustice/training/apiasejissues.html).

40. Lester et al., *Environmental.*

41. Lester et al., *Environmental.*

42. Unlike the civil rights movement, the environmental justice movement seems to be dominated by women, many of them older than the college and high-school students of the 1960s movements. Many of these women bring skills and organizing experience from previous movements and from the black church.

43. Anna Holden and Charles Simmons, "Community Health and Environmental Justice: Burning Issues in Detroit," *Everyone's Backyard* 17, no. 3 (Fall 1999).

44. We must change our production patterns to be consistent with the Earth's life cycle by using chemicals and materials that are more linear in character. For example, products and/or waste that will be around for hundreds and even thousands of years need to be abandoned in favor of products that will biodegrade more quickly.

2

Environmental Justice and the Social Determinants of Health

Virginia Ashby Sharpe

In his *Nicomachean Ethics,* Aristotle provides us with one of the earliest and most influential articulations of justice in the philosophical tradition. In its most general sense, says Aristotle, justice means that equals are treated equally and unequals may be treated unequally.[1] How this very abstract principle has been filled in and translated into public policy is nothing less than the history of political philosophy, the history of economics, and the ongoing battleground of local, national, and international politics. In thinking with Aristotle about justice, it is important to keep in mind that the question of justice and, therefore, the question of politics are fundamentally *ethical* questions about what constitutes a good, healthy, flourishing society. If we claim to care about justice, we are always ultimately accountable with respect to this ethical question.

In this chapter, I draw on empirical evidence that has been collected over the past thirty years that provides a rich basis for answering the question of what constitutes a good society. I use this evidence to break what I believe is a stalemate in the literature on environmental justice—a literature that has often been narrowly focused on empirical evidence for the *biophysiological* effects of toxic exposure and the *intentions* behind noxious industry siting in minority and low-income communities. Specifically, I draw on broader empirical evidence from Nobel laureate Amartya Sen and health and development economists and epidemiologists around the world who have determined that countries with the least disparities in income are the countries with the highest life expectancies. In other words, the degree of income equality in a society has a significant impact on the social cohesion of that society, which is, in turn, one of the most powerful determinants of health.[2] Put even more simply, it is not the wealthiest countries that have the best health, it is the countries that are the most egalitarian.[3] I make use of some of this evidence to shed light on the goals of environmental justice.

As a starting point, I clarify Aristotle's formal principle of justice, that "equals be treated equally and unequals unequally," by addressing what it might mean to say that X and Y are unequal. At the end of the chapter, I examine what it means to *treat* unequals unequally in the achievement of environmental justice.

WHAT DOES IT MEAN TO SAY THAT X AND Y ARE "UNEQUAL"?

What does it mean to say that X and Y are not equal? It might mean one of two things, which we can distinguish as inequality in the *absolute* sense or inequality in the *relative* sense.

Absolute Inequality

If we say that "X is not equal to Y" in the *absolute* sense, we are making a judgment that Y is somehow superior to X because Y has some characteristic that X does not have. Throughout human history, we have drawn boundary lines around the moral sphere or the sphere of moral consideration. Those inside the sphere are considered to be equal by virtue of some shared characteristic, such as having a soul (which was Descartes's basis for excluding animals), having white skin and male gender (which was the U.S. Constitution's basis for excluding people of color and white women from full citizenship), or owning property (this has been the basis for excluding non–property owners from voting and other forms of political participation). Those lacking particular qualifications are seen as morally inferior and therefore not entitled to equal treatment or consideration.

Of course, one of the great moral triumphs of liberal democracy is its premise that human dignity and self-determination are by themselves a sufficient basis for inclusion in the moral sphere. The antidote to the exclusion of people based on arbitrary characteristics of race, class, gender, religion, or sexual orientation is the doctrine that all humans are morally equal by virtue of their dignity and autonomy. This philosophical conception of moral equality has been the basis for the guarantee of civil and political rights in the United States and around the world. It is also the basis on which we judge actions and policies to be racist, classist, sexist, or discriminatory.

In the context of environmental injustice, it is possible to read particular policies as precisely this form of exclusionary bigotry, that is, bigotry based on a belief in the *absolute* inequality of the excluded group. For example, as reported in the March 15, 2001, *New York Times,* the New York State Power Authority and the George Pataki administration have acknowledged that the electric power generators that it wants to install around New York City would go into poor, heavily minority-populated communities of the Bronx, Queens, and Brooklyn.[4] Given the history of marginalization and exploitation of racial, ethnic, and economic minorities, it is not unreasonable to see this as a further example of racism and active prejudice. When it looks like something we call "progress" is built on the backs of the poor and people of color, it is very plausible to assume that there is discrimination at work— discrimination based on the notion that one group is *absolutely* inferior to another and, therefore, not deserving of equal treatment.

Anyone who has written or read about environmental racism knows there is presently a debate about whether certain land uses, for example, really *are* evidence of racism. The debate essentially comes down to whether it can be proven that the siting of locally unwanted land uses in minority communities is based on a belief in the absolute inequality, the moral inferiority of those minorities and, therefore, intentional discrimination against them. This is the basis of equal protection claims (which appeal to the equal protection clause in the Fourteenth Amendment to the U.S. Constitution) in environmental justice lawsuits. A valid equal protection claim requires proof that the defendant *intended* to discriminate against the plaintiff. The best example of this debate is: "which came first, the dump or the minority community?" The argument is that if the dump came first and the minorities moved in afterward on the basis of affordability, this does not constitute racism or discrimination because these people made a voluntary choice to move into the neighborhood. If the siting of a dump is challenged only on the basis of discriminatory intent, then disparities caused by market forces such as residential mobility are immune from redress under the equal protection clause.[5]

When I promised that I would draw on evidence to break a stalemate in the environmental justice policy literature, this is one of the areas of impasse I had in mind. The data that I provide on what is known as "capability deprivation"[6] reveal that injustice is something that arises from economic and political factors more subtle and far-reaching than whether a particular land use is intentionally racist. One of the obvious problems with the argument that the location of a dump in a minority community is not racist because minorities have voluntarily chosen to move or stay there is that it assumes that "residential mobility" and the "dynamics of the housing market"[7] are somehow benign or morally neutral. Understanding the afflictions of inequality means recognizing the harms associated not only with intentional discrimination or the immediate and proximate dangers of toxic materials, but also with the "toxicity of social circumstances and patterns of social organization."[8]

Before I discuss capability deprivation and what it might ultimately mean to have a good, just, and flourishing society, I need to briefly return to Aristotle's notion that justice means that equals should be treated equally and unequals may be treated unequally. I started out by asking what it might mean to say that two people aren't equal. What I just described is *absolute* inequality—the belief that someone or some group is irredeemably unequal because they lack some necessary qualification. Racism, sexism and economic discrimination are understood as legacies of this belief.

I also pointed out that the premise of liberal democracy—that all people are created equal and are moral equals solely by virtue of their humanity and human dignity—was the Enlightenment's remarkable antidote to this view. The challenge for liberal democratic theory since has been to figure out the relationship between our moral equality and the very real, substantive differences between us. And this brings us to the notion of *relative* (rather than absolute) inequality.

Relative Inequality

To say that two people are not equal *relatively speaking* is to say that they are differently situated. The key question is *which* differences of situation are morally important; that is, which inequalities of health, income, material circumstances, opportunity, gender, ethnicity, or age are inequitable or unfair and which are not. In our country, Medicare and Medicaid, which are intended to assure health-care access to the elderly and the poor, respectively; the Americans with Disabilities Act (ADA), which is intended to assure equal access and opportunity to the disabled; and affirmative action, which is intended to assure equal citizenship, access, and opportunity to those historically disenfranchised from political, civil, and economic participation[9] are all policies that seek to remedy relative inequalities seen to be unfair and inequitable. The ADA, for example, indicates that we don't believe it's fair to be excluded from access to a job for which a person is qualified simply because that person cannot walk up the stairs to the office. The pressure to abandon these policies, to curtail affirmative action, and to cut back Medicaid are, by contrast, based on the belief that inequalities in income and opportunity are *not* inequitable and therefore require no remedy. This view is best reflected in the libertarian axiom that such inequalities may be "unfortunate but they are not unfair."[10]

At this point, it will be helpful to take a closer look at libertarian political philosophy, because it is the philosophy that undergirds the free market and one that has guided recent Republican presidential administrations in this country. Taking the Enlightenment belief in the moral equality of all people to its extreme, free-market libertarians believe that there are *no* morally relevant differences between people at all. Everyone is the same by virtue of their liberty and this fact is best respected by making sure that there is no infringement on the liberty of another without his or her consent.

This view shows up in the environmental justice literature in a number of ways. First, free-market libertarians argue that those who move into areas environmentally degraded by, for example, dumps or toxic sites do so voluntarily, that is, freely. If, as libertarians believe, justice is freedom and freedom is not violated, then there is no injustice. Second, libertarians argue that if people freely accept compensation for bearing particular burdens—such as living near an undesirable land use or selling their toxic home or land to a polluting company—then their liberty has been respected and, again, there has been no injustice done. Third, according to free-market libertarians, one is free to do anything as long as it does not harm someone without his or her consent. Giving absolute priority to the principle of liberty in this way places the all-important burden of proof on the aggrieved parties to prove that they are at risk or that they have been involuntarily harmed. The issue of the burden of proof is one of the most important battlegrounds in the debate over environmental justice.

Following this libertarian approach, a free-market, antiregulatory philosophy favors placing the burden on potential claimants to prove harm. The rationale here is, again, that liberty is paramount and restrictions to liberty constitute unjust interference. In the case of a siting decision, for example, a company subscribing to this philosophy would assert that its liberty must be unrestrained—its activities unregulated—unless it can be proved that its activities are dangerous. Environmental justice advocates including Robert Bullard, Bunyan Bryant, and others[11] articulate a precautionary principle that essentially shifts the burden of proof, requiring that the proponents of an activity (such as a power-generating plant, an incinerator, a hazardous waste facility, or a nuclear power plant) prove that it is safe or will not cause significant harm. The obstacles to proving either harm or safety are enormous and include determining the scope of the impact(s) (who or what will be affected: children, adults, property, animals, ecosystems, future generations); the nature and magnitude of impact(s) (great or minimal? causing disease, death, or reduced quality of life?) the scale of the impact(s) (short term? long term?); the probability of the impact(s) (likelihood?); the acceptable level of harm or safety (value trade-offs such as jobs versus a healthy environment). Those who argue for a more precautionary approach believe that given what we know about the hazards of modern industrial life, it no longer makes sense that the enormous obstacles to proof should be borne by the public—a fact only underscored by the unequal distribution of burdens. Supporters of a preventive or precautionary approach also argue that the current burden of proof has provided an opportunity for infinite delay as adversaries spend years contesting the meaning of data. As Bunyan Bryant has argued, "causality arguments or issues of certainty are often used to rationalize inaction, particularly when it has been economically or politically expedient to do so."[12] We have ample evidence of such delays in the debate over dredging PCBs from the Hudson River, the federal government's dioxin reassessment, and the harms associated with cigarette smoking. Though I believe that we as a nation should move toward the adoption of a more precautionary approach, I also believe that the debate about the burden of proof as well as many of the ongoing debates over the empirical evidence linking exposure to adverse health effects represent an impasse in the environmental justice literature. I am not suggesting that risk assessment is unimportant or that a broader public policy conversation about the burden of proof should not go forward. What I *am* suggesting is that we need a new perspective on these debates to move us toward the achievement of a more just and healthy society, one that understands "the environment" and "environmental risk factors" to include more than potential biophysiological insults. In other words, in order to have a more robust understanding of "environmental justice" we need to look at the conditions "upstream" that broadly determine the health of society.

In the service of this broader conception of "the environment," "environmental risk factors," and "environmental health," I would like to make good on my promise to share some data on what are called "the social determinants of health"—data that challenge the prevailing assumption of an inherent tension between "progress" and social justice and between economic efficiency and equity, data that help us to understand just what is at stake in relative inequality.

THE SOCIAL DETERMINANTS OF HEALTH

The 1974 World Population Conference in Bucharest, Romania, represented a turning point in the global understanding of the relationship between population and development. For decades, the working assumption of Western development experts was that population growth was the cause of underdevelopment. On the basis of this assumption, national and international policy promoted population control initiatives, sometimes coercive ones, as the means to economic growth and progress. At the conference, third world delegations turned this assumption on its head. Population growth, they argued, was not the *cause* of underdevelopment, it was its *consequence*.[13] Empirical evidence from Kerala, India, for example, gave rise to the view that underdevelopment was the result of systemic inequalities that in turn led to uncontrolled population growth. The evidence established that population growth would be reliably curbed only by the eradication of infant mortality, inequities in income, improved education, and economic participation. In Kerala, health-care and nutrition initiatives, property ownership by women, and programs to eliminate illiteracy, especially among women, resulted in zero population growth. In other words, the increase in social and political participation by women meant that they were able to develop capabilities other than childbearing as a means of social participation. The evidence from Kerala indicated that the enhancement of social justice for women and children was the best means by which to slow population growth and to enhance overall life expectancy.[14]

Since then, evidence has mounted that just as social factors such as gender equity, literacy, and maternal and child health services influence the rate of reproduction, so too do social factors influence the life expectancy of groups both within and between societies. More specifically, the evidence suggests that greater equity in society also produces greater overall health.

For more than a century, epidemiological studies—that is, studies of health and disease in populations rather than in individuals—have shown that differences in social position are closely related to health and longevity within societies. For example, in a developed country the wealthy may have a life expectancy two to four times higher than that of the poor.[15] A conclusion long drawn from these data was that the wealthier the person or society, the healthier it would be. If we were to chart this conclusion on a graph, we would show that as income increased, so too would life expectancy (fig. 2.1). This conclusion has, in part, provided support for the argument that increases in income alone will improve health status. It was not until international comparisons were conducted that we began to understand that the relationship between health and income is more complex.

The 1998 United Nations *Human Development Report* showed that once a country passes a threshold level of income, its entire population "can be more

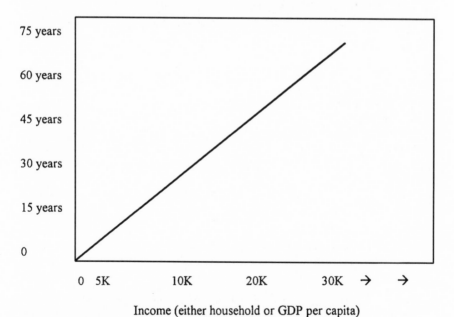

Income (either household or GDP per capita)

Figure 2.1. The Conventional Wisdom about the Relationship between Wealth and Health (Life Expectancy)

than twice as rich as another without being any healthier"[16] (fig. 2.2). The difference in per capita GDP (gross domestic product) between the United States and Costa Rica, for example, is roughly $21,000. However, Costa Rica's life expectancy exceeds that of the United States. Likewise, Cuba has a life expectancy on par with that of the United States but a per capita GDP roughly one fifth of that of the United States. These and similar data, amassed over the past thirty years by eight research groups in ten different data sets, indicate that something other than wealth produces relatively healthier societies and something other than poverty produces relative illness. Cross-national comparisons show that health disparities are correlated with relative disparities in socioeconomic status. The most significant factor in the achievement of health in a society is *the size of the gap between its rich and poor members*. The wider the gap, the lower the overall life expectancy. The narrower the gap, the higher the life expectancy. In some of the most striking research, an analysis of the relationship between the poorest 20 percent of the population and the richest 5 percent indicated that the higher the incomes of the richest 5 percent, that is, the wider the income differences within a society, the higher the infant mortality rates were for the society overall.[17] The healthiest societies, by contrast, were the most egalitarian societies.

Additional research has shed light on the specific ways in which social inequalities adversely affect the overall health of a society by creating and systematically depriving an underclass. In the United States, "the most inegalitarian states with respect to income distribution invest less in public

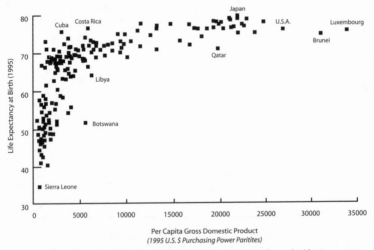

Figure 2.2.   The Relationship between Country Wealth and Life Expectancy
Source: N. Daniels, B. Kennedy, and I. Kawachi, "Why Justice Is Good for Our Health: The Social Determinants of Health Inequalities," Daedalus 128, no. 4 (1999): 215–52; United Nations Human Development Report Statistics, 1998.

education, have larger uninsured populations, and spend less on social safety nets."[18] Similarly, as figure 2.3 shows, homicide rates among U.S. states correlate with the percentage of total household income received by the least well-off 50 percent. In 1990, Louisiana and Mississippi—the states where the poor receive the smallest share of its state's total household income—had the highest homicide rates. By contrast, the lowest homicide rates occur in states with the most egalitarian distribution of income.

What these data indicate is that in societies where basic needs of food, clothing, shelter, and sanitation have been met, health differences are ex-

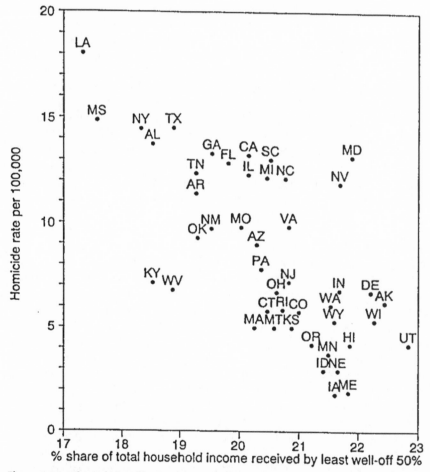

Figure 2.3. The Relationship Between Income Distribution and Homicide among the States of the USA in 1990
Source: R. Wilkinson, *Unhealthy Societies: The Afflictions of Inequality* (New York: Routledge, 1996), 157.

plained by relative deprivation within the society. Since basic needs have been met, this deprivation refers not to material necessities but to what Amartya Sen has called "capabilities," or the various political, economic, and social freedoms that enable people to lead lives that they have reason to value. These deprivations of the underclass reinforce and promote a loss of social cohesion and undermine the overall health of society.

It might be useful at this point to provide an example not from social science but from literature. In his brilliant 1940 novel, *Native Son*,[19] Richard Wright paints a searing picture of the destructive force of racism and classism through the tragic intersection of the lives of Bigger Thomas, a young, poor black man, and the Daltons, a wealthy white family. At the beginning of the book, we see Bigger, his mother, and two siblings waking up in their one-room, rat-infested apartment. The apartment is in Chicago's "Black Belt"— the only area of the city where white real estate owners will rent to black people. Bigger has a job interview that night with Mr. Dalton, and his mother's anxiety makes it clear that the well-being of the family depends on Bigger's taking the job. Mr. and Mrs. Dalton are privileged by wealth, race, and social position. They see it as their obligation to help the less fortunate to make something of themselves. They offer Bigger a job as their chauffeur, replacing another black man whom they had helped put through night school. Mary, the Dalton's daughter, a spoiled and naive but well-meaning college student, believes that communism will resolve racial and economic divisions. Without understanding Bigger's vulnerability as a black man in a white society and as a poor black employee to a wealthy family, Mary flouts social taboos and insists on treating Bigger as if there were no differences between them, sitting in the front seat of the car with him, making him eat with her and her boyfriend at a restaurant in Bigger's neighborhood. Through a series of events, Bigger's fear, shame, confusion, and marginalization lead to the death of Mary Dalton. Her death not only results in the disintegration of Bigger's life but also creates a frenzy of fear, distrust, and social disruption in Chicago. The tragic irony of the story lies in the fact that Mr. Dalton, though he extends his charity to Bigger, is also the owner of the South Side Development Company, the real estate company from which the Thomases rent their squalid apartment for eight dollars a month—more than whites are charged for comparable housing. When, at the coroner's inquest, Mr. Dalton is asked why he doesn't simply charge blacks less rent rather than compensating some with charity, he says that it would be "unethical" to undersell his competitors.[20] When he is further asked whether he thinks conditions under which the Thomas family lived in his rental apartment might in some way have contributed to the death of his daughter, he says "I don't know what you mean."[21] In his final statement to the court hearing Bigger's case, Bigger's lawyer reiterates the upstream conditions of American society in 1940 that culminated in the death of Mary Dalton. "Taken collectively, [American

Negroes] are not simply twelve million people; in reality they constitute a separate nation, stunted, stripped, and held captive *within* this nation, devoid of political, social, economic, and property rights."[22] Those who today argue that market forces such as residential mobility bring poor residents voluntarily to environmentally degraded areas would do well to reflect on the ways in which today's market conditions similarly force particular housing choices.

The civil rights movement and subsequent legislation from the 1960s and 1970s has provided the basis for equal opportunity for people of color, the poor, and women in the United States. During this same period, it has been argued that many of the burdens of industrial development—the so-called externalities of waste, noise, odor, poor housing, fragmented neighborhoods, and illness—have fallen disproportionately on these groups. Christopher Foreman and others have argued that studies offered to substantiate claims of disproportionate environmental impacts on minority and low-income communities are highly questionable, having severe methodological and evidentiary limitations.[23] Foreman disputes any proven causal relationship between industrial activities and adverse health effects on minority communities, except for lead exposure in minority children and chemical exposure among farmworkers. However, he does acknowledge that "health," understood in the more inclusive sense of "quality of life," may better describe the aspirations of those fighting for environmental justice.[24] Foreman's distinction between health and quality of life is important because it highlights a pervasive and often implicit assumption that "health" is reducible to the absence of disease. This definition of health is a by-product of our medical paradigm. If we think about health within the broader public health and development paradigms, however, we get a much richer conception of what it would mean to be healthy, both as a society and as an individual.

Health and Capabilities

From the perspective of public health and development, health is not simply the absence of disease; rather, it is the use of one's human capabilities to lead a life that one has reason to value.[25] Evidence from the public health and development literatures indicates that psychosocial factors such as political inclusion, opportunities for utilizing economic resources, stress, and literacy are the major determinants of health, both for individuals and for societies. In other words, the most significant risk factors for illness and premature death are *social* in nature, with poorer health outcomes closely linked to inequalities in the distribution of wealth and opportunity.

What, then, does this have to do with environmental justice? First, this analysis of the social determinants of health tells us that the causal story of environmental health risks is not reducible to an empirically established linkage between exposure to chemical X and adverse health effects Y and Z. To know that racial and economic minorities are burdened by environmental inequities is to know that they are disproportionately subject to adverse social conditions that limit quality of life and human capabilities. This would certainly include living and working in environments degraded by crime, industrial siting, or pollution, or compromised by political exclusion. These data also tell us that "environment" means more than the air we breathe, the water we drink, and the ground in which our food is grown. "Environment" also means the social environment in which we "live, work, play."[26]

Based on his research on the social determinants of health, Richard Wilkinson sums this up, saying that "it is the social feelings that matter, not exposure to a supposedly toxic material environment. The material environment is merely the indelible mark . . . of one's social exclusion and devaluation as a human being."[27]

CONCLUSION—TREATING UNEQUALS UNEQUALLY

I began this paper with reference to Aristotle's notion that a just society is one where equals are treated equally and where unequals may be treated unequally. We are now in a position to understand what this might mean. In Aristotle's day, slavery was an unquestioned part of the social order and women were not granted citizenship. For Aristotle, therefore, there were people who were unequal in the *absolute* sense by virtue of their ethnicity, gender, and socioeconomic status. In modern democratic societies, we have rejected the idea of the absolute inequality of people. The premise of American democracy is that all people are created equal. This fact about our *moral* equality provides a basis on which to make sense of our *relative* inequality—those very real material and historical factors that situate us differently. The data that I have referred to on the social determinants of health indicate that it is not the wealthiest countries but the countries that have the smallest income differences between rich and poor that have the best health. Inegalitarian societies, those with larger relative income differences are less socially cohesive and suffer more of the corrosive effects of social and economic divisions. In this sense, relative inequality adversely affects not only the disadvantaged, but also the health of society as a whole. For Aristotle, the question of justice is ultimately a question about what constitutes a good, healthy, flourishing society. Given the data that we have examined, it seems that the most pressing problem in and between societies is the alleviation of relative inequalities. Paradoxically, this will require treating unequals unequally—though in John Rawls's[28] sense of seeking policies that achieve the greater advantage of the *least* well off. In other words, without a redistribution of income and welfare advantages, including social and environmental benefits to the least well off, society as a whole will not flourish.

I conclude with three final observations. First, the evidence on the social determinants of health suggests that our habit of pitting justice (equity) and progress (growth, development, efficiency) against one another is a false dichotomy. Efficiency and growth are not ends in themselves; they are *means* to ends. The benefits of economic growth are not direct; they are indirect. The challenge for society, therefore, is the conversion of income into human development. These data suggest that efficient growth is achieved when there is greater equity. Second, assessing the harmful effects of environmental agents should be based on the notion of health that derives from public health, not from medicine. This will require incorporating variables from the UN Human Development Index into our analyses of risk and harm.[29] Third, the orientation of policy makers should not simply be backward-looking— involving compensation based on past discrimination (absolute inequality)—but forward-looking, to the type of society that authenticates social cohesion and participation. Only if we take seriously the dramatically unsustainable effects of relative inequality will we be able to move forward together.

From this broader perspective, environmental justice can be seen as a movement for social justice that recognizes the deprivations that are associated with environments degraded not only by specific pollutants, but also by social and political factors. Environmental justice can also be seen as a new form of environmentalism, one that, as Robert Gottlieb observes, is "democratic and inclusive, an environmentalism of equity and social justice, an environmentalism of linked natural and human environments, an environment of transformation."[30]

NOTES

1. Aristotle, *Nicomachean Ethics,* trans. Terrence Irwin (Indianapolis: Hackett, 1985), 1131a, 20–30.

2. R. Wilkinson, *Unhealthy Societies: The Afflictions of Inequality* (New York: Routledge, 1996), 5.

3. Wilkinson, *Unhealthy Societies,* 246.

4. R. Pérez-Peña, "State Admits Plants Headed to Poor Areas," *New York Times,* March 15, 2001, B1.

5. V. Been, "What's Fairness Got to Do with It? Environmental Justice and the Siting of Locally Undesirable Land Uses," *Cornell Law Review* 78 (1993): 1001–1085, 1014.

6. A. Sen, *Development as Freedom* (New York: Anchor, 1999), 20.

7. C. Boerner and T. Lambert, "Environmental Justice?" *Center for the Study of American Business Policy Study Number* 121 (April): 16, (St. Louis, Mo.: Washington University, 1994).

8. Wilkinson, *Unequal Societies*, 23.

9. R. Dorkin, "Race and the Uses of Law," *New York Times,* April 14, 2001, A17.

10. R. Nozick, *Anarchy, State and Utopia* (New York: Basic, 1974).

11. B. Bryant, "Issue and Potential Policies and Solutions for Environmental Justice: An Overview," in *Environmental Justice: Issues, Policies, and Solutions,* ed. B. Bryant, 8–34 (Washington, D.C.: Island, 1995); R. Bullard, "Environmental Justice for All," in *Unequal Protection: Environmental Justice and Communities of Color,* ed. Robert Bullard (San Francisco: Sierra Club, 1994), 3–22; Earthcharter, www.earthcharter.org/draft/charter.htm; S. Steingraber, *Living Downstream* (New York, Vintage, 1997), 284.

12. B. Bryant, "Issue and Potential Policies," 10.

13. M. Conroy, K. Kelleher, R. Villamizar, "The Role of Population Growth in Third World Theories of Underdevelopment," in *Ethical Issues of Population Aid: Culture, Economics and International Assistance,* ed. D. Callahan and P. G. Clark, 171–206 (New York: Irvington, 1981).

14. J. Ratcliffe, "Poverty, Politics and Fertility: The Anomaly of Kerala," *Hastings Center Report* 7, no. 1 (1977): 34–42; Sen, *Development as Freedom,* 219–23.

15. N. Daniels, B. Kennedy, and I. Kawachi, "Why Justice Is Good for Our Health: The Social Determinants of Health Inequalities," *Daedalus* 128, no. 4 (1999): 215; Wilkinson, *Unequal Societies,* 215.

16. Wilkinson, *Unequal Societies,* 3.

17. Wilkinson, *Unequal Societies,* 83.

18. Daniels, "Why Justice is Good for Our Health?" 223.

19. R. Wright, *Native Son* (New York: Harper & Brothers, 1940).

20. Wright, *Native Son,* 278.

21. Wright, *Native Son,* 279.

22. Wright, *Native Son,* 333.

23. C. Foreman Jr., *The Promise and Peril of Environmental Justice* (Washington, D.C.: Brookings Institution, 1998).

24. Foreman, *Promise and Peril,* 88.

25. Sen, *Development as Freedom.*

26. R. Gottlieb, *Forcing the Spring: The Transformation of the American Environmental Movement* (Washington, D.C.: Island, 1993), 5.

27. Wilkinson, *Unequal Societies,* 215.

28. J. Rawls, *A Theory of Justice* (Cambridge, Mass.: Harvard University Press, 1971).

29. United Nations Development Programme, *Human Development Report 1990* (New York: Oxford University Press, 1990), and subsequent yearly reports. <http://www.undp.org/hdro/report.html>

30. Gottlieb, *Forcing the Spring,* 320.

3

Green Imperialism: Indigenous People and Conservation of Natural Environments

Manuel Lizarralde

> You people of the developed nations have destroyed your surrounding natural areas and have too many children. Stop having children and let other societies develop too. The earth is small and we need to coexist on this planet. We, the Yanomami people of Brazil and Venezuela, still have our forest and animals. We are few and need to have more children to survive as an ethnic group. The land does not belong to any specific individuals. We do not have the right to posses the land and destroy it. It belongs to all of us and we all need to take care of it for them and each other. If we do this, we will survive to the next millennium.
>
> *David Kopenawa, Yanomami leader, April 7, 2002*

The goal of this chapter is to explore the notion of a new form of imperialism, green imperialism. Green imperialism is based on the often unrealistic preconception that indigenous people will happily accept the role of protecting and conserving their environment at all costs, while at the same time human societies in the developed nations (which account for 21 percent of the global population) continue their overconsumption of natural resources.[1] Under the concept of the "Noble Savage," indigenous people are expected to continue conserving biodiversity with the assumption that they do not want access to even the most basic of Western commodities. This notion of the indigenous people's way of life is a form of slavery in many instances. In chapter 1, Bunyan Bryan provides us with a history and a description of the national form of environmental racism and discrimination. Green imperialism is another form of environmental discrimination, focusing on contemporary and future environmental justice issues in the third world and on indigenous people. This chapter focuses on people of the tropical humid regions and explores in particular the case of the Barí people, an indigenous society of Venezuela. This focus provides many appropriate examples of how green imperialism is affecting indigenous people and developing nations alike.

The concept of green imperialism is rooted in the notion that there is both a large discrepancy between living standards of developed nations and third world indigenous peoples and an expectation of who is subsidizing the conservation of natural environments. This discrepancy is a product of European colonization and the appropriation of rich resources from other nations in the world, including indigenous ones, over the past five hundred years. Today, most of the world's indigenous people, representing 5 percent of global population, live on 19 percent of the land that is rich in plant and animal species not found anywhere else on the planet.[2] Although it has been empirically proven that indigenous people have been living in a relatively harmonious relationship with their environment and have been causing minor detrimental impacts on their environment, it does not mean that their way of life will continue indefinitely.[3] This low-technology, subsistence way of life has its sacrifices. It is hard work to produce the needed resources with the traditional technologies indigenous people use and, as a result, there is often high mortality and morbidity. It is unlikely that most indigenous people will keep their way of life to conserve the natural environments of the world, even though many of us would like to believe indigenous people will continue to do this. Unfortunately, many indigenous people are rapidly changing their traditional ways of life and adopting more comfortable, westernized existences. Throughout different human societies and in human history, the goal of most cultures is to find a way to lead an easy life, which is not equivalent to leading an ecologically harmonious life. This is one reason human technology has been constantly improving our tools and forms of shelter since the times of the first hominids.

The definition of the "Noble Savage" and "Noble Indian" has been highly debated in the literature. The "Noble Savage" or "Noble Indian" represents an assumption that indigenous societies live romantic lives that are harmonious with nature. There is the dangerous tendency among anthropologists, in their assumptions of unchanging indigenous societies, to describe these societies as idealized, noble societies. It is clear that there are cases of sustainable societies, but their numbers are clearly diminishing. The problem is that this notion is used without an understanding of the ecological, historical, social, and political context within which a given society is framed. On the one hand, we have many examples of ecologically sustainable societies at different places and times. On the other hand, we have observed arguments that other societies have been destructive and unsustainable. The intention of this chapter is to clarify the complexity of this notion of the "Noble Savage" and its negative impact on indigenous societies in the present and on the future conservation of natural areas where these human populations are found.

THE NOBLE INDIGENOUS PEOPLE

This chapter does not intend to disprove the existence of indigenous people who are conservationists. On the contrary, in the literature we find many examples of indigenous societies with a deep respect for their natural environment.[4] Moreover, many authors have demonstrated the nature of indigenous peoples as conservators in the past and present. This is clearly stated by Stan Stevens: "Indigenous Peoples' knowledge, conservation beliefs and values, environmentally adaptive and sensitive land use, resource management practices, and determined defense of territory and natural resources have enabled many of them to inhabit their homelands for centuries without devastating their ecosystems and biodiversity."[5] The problem, as stated by Marianne Schmink, Kent H. Redford, and Christine Padoch is that "[t]his assumption that traditional peoples will remain 'ecological noble savages' . . . is not only incorrect but potentially disastrous to the people themselves."[6] We are dealing with a rapidly changing world that is clearly affecting the conservation of resources.[7]

In his book, *The Ecological Indian,*[8] Shepard Krech III provides a complex analysis of what makes conservation possible and why indigenous people who choose development for economic reasons are viewed as not being conservative. Krech does not accept openly the notion of the "Noble Indians," but he does not completely disprove it either. Krech shows many examples and agrees with many writers, like the historian William Cronon (*Changes in the Land,* 1983), that indigenous people were in most cases conservators of their environment in New England. If they were not, it was because changes imposed by European descendants and westernization affected their nature and their needs.

Krech quotes Dennis Martinez, an O'odham and Chicano writer who comments that Native Americans have "taken care of the landscape for thousands of years."[9] They are "wise environmental managers," as Martinez remarks, who understand "ecology and land stewardship."[10] However, Krech states that "the debate" over significant natural resource holdings "is pretty much over as to whether we should engage in economic development." As far as he and many others are concerned, such development is the only course for the future."[11] Additionally, in his concluding paragraph, Krech states,

> Yet most underscore the complexities involved in the decisions on natural resource and land issues. Many people in Indian Country desire the trappings of

middle-class American life—cars, televisions, stereos, jobs, money—but do not want to lose their Indianness or sense of belonging to place. As one Choctaw (who reminded his non-Indian interviewer that he did not want "to be you") remarked, "I like living in this community, and I like being Choctaw, but that's all there is to it. Just because I don't want to be a white man doesn't mean I want to be some kind of mystical Indian either. Just a real human being.[12]

This voice states the ambiguity of the construction of "Noble Savage" society and the need to live the westernized lifestyle. Krech's example of the complexity of the notion of the "Noble Savage" is from the United States. The following section will explore this issue in the rainforest and its implication in relation to rainforest conservation.

PRESSURES FOR THE CONSERVATION
OF AN AREA: THE TROPICAL RAINFOREST

One particular scenario of revitalizing and preserving the image of the "Noble Savage" is based on preconceived notions of indigenous people in rainforests around the world. The rainforest has been extensively destroyed[13] in the past fifty years, alerting almost all societies in the world to the ecological implications, even societies within the rainforest. Parallel to this destruction, many indigenous people who live in these forested regions have become partially westernized.[14] In order to satisfy increasing monetary needs, indigenous people have also often taken active roles in the destruction of their forests. The motive behind this role is development and raising the standard of living of a growing world population.[15] Now, especially with the initiative of the developed nations, there is an urgent need to stop the destruction of the remaining 46 percent of the rainforest and save the indigenous cultures living in it.

One major problem is that there are several incorrect assumptions about rainforest societies. Members of these traditional societies, far from fitting the romantic notion of the "Noble Savage," also have the desire to develop a more comfortable way of life, like ours, with all of the commodities available in the market. Marshall Sahlins's "Notes on the Original Affluent Society"[16] is an example of the literature expressing the notion of how easy is the life of traditional societies, with their simple technology.[17] But this notion does not explain why many indigenous societies are not keeping their traditional ways of life but rather are shifting to more westernized ones. The answer is clearly stated by John H. Bodley: "Certainly, not all indigenous peoples accept the idealized view of their traditional culture, nor do they necessarily all support the self-determination movement. Many individuals may find the personal rewards potentially available in the dominant society to be more attractive than the traditional tribal life style."[18] Bodley's statements seem to reflect the common path many indigenous people are taking around the world. These problems have been observed in my fieldwork with groups such as the Barí people in Venezuela and the Machiguenga people in Peru.

It is true that some societies and individuals living this traditional subsistence life are happy and healthy as long as (1) their population density is very low, (2) their culture maintains or has developed subsistence strategies of sustainable use of their environment, and (3) resources are abundant.[19] There are indigenous populations in the rainforest who live self-sufficient, sustainable lives in the present and would like to keep it that way, for example, the Hoti people of Venezuela.[20] According to Egleé Lopez Zent and Stanford Zent, the Hoti people carefully utilize resources like wild honeybees and palms so as not to deplete them.[21] Unfortunately, such societies are scarce in numbers and are rapidly decreasing.[22] The majority have chosen to be consumers of western goods and have gradually adopted partially westernized lives for convenience. Since their population densities have dramatically increased due to reduction of their territories and increased population, their traditional subsistence practices have not provided sufficient resources.

The basic irony is that we want these traditional rainforest inhabitants to preserve their sustainable ways (with the assumption that little or no destruction of their environment is occurring), while they want our way of life because it is safer and more commodious. Generally speaking, mortality rates in traditional societies are high—commonly as many as half of their children die before reaching the age of five. With Western medicine, children in these societies tend to have a higher chance of survival. Moreover, westernization offers the labor-saving power of our technology, from steel knives to motor vehicles. All of these obviously cost money, meaning indigenous people have to sell resources (generally lumber, meat from wild animals, or cash crops), which often promotes forest destruction.

It is unrealistic and irresponsible to assume that indigenous people should continue their risky and biologically stressful lives filled with demanding physical effort, while we continue our luxurious but unsustainable lives that consume twenty to fifty times more resources per capita than their "traditional ways."[23] This is the essence of green imperialism, in which Westerners, practicing hyperconsumption of resources, expect rainforest inhabitants to live a way of life that consumes relatively little. They are expected to conserve biodiversity for us and maintain traditional precontact subsistence strategies even though ecological and economic conditions around them have changed. To reverse the destructive process currently playing out in the rainforest, there are some questions we need to ask. What are the alternatives for the societies living there? Since maintaining the traditional sustainable, "affluent" indigenous societies for the next millennium is far from realistic, how can current indigenous people living in the rainforest survive current trends? How are we going to stop the loss of biodiversity and adapt in the future under these changing socioeconomic, demographic, and ecological circumstances?

It is difficult to project and anticipate the exact problems indigenous people will face in the future. There is no doubt that we have to acknowledge the importance of the indigenous people's participation in protecting the forest, since they have the ultimate right to do so.[24] Most of the forest is, after all, in their territory or in areas that were inhabited by indigenous people for a long time. However, we also have to acknowledge that the conditions of indigenous people are unstable and that many changes are affecting their relationships to the environment. Bodley points out that there is a naive assumption that the culture of indigenous people "is or should be static."[25] With exposure to the Western world, changes tend to come with incredible speed, especially with the introduction of Western medicine and the global market economy.

In recent years, extensive literature has appeared on the survival of indigenous society and the conservation of the forest from many perspectives, ranging from those of scientists to those of indigenous people.[26] In addition, Jason Clay has articulated these ideas in various articles and books.[27] The purpose of this chapter is to present a case that illustrates the main problem indigenous people are facing in relation to saving their natural environment in the next millennium.[28] The case of the Barí indigenous people illustrates the predicament of both the use of the rainforest and its conservation in a rapidly changing world.

THE CASE OF THE BARÍ PEOPLE

The Barí are an indigenous society living in the northwestern part of Venezuela and across the border in Colombia. They are slash-and-burn horticulturalists who traditionally lived semisedentary communal lives, supplementing their diet with fishing and hunting. They moved from one longhouse to another, depending on the availability of garden and forest resources. They were first contacted by outsiders in 1960 and quickly experienced many changes, including significant loss of territory and their traditional way of life. They shifted to a more sedentary way of life because their territory was vastly reduced and they had to settle permanently in order to prevent ranchers and peasants from claiming the vacant lands should they leave. By becoming sedentary, they could no longer use forest resources in various areas sparingly and sustainably, but resorted to more intensive use in a single area. The depletion of natural resources forced the Barí to abandon their traditional subsistence activities and enter the market economy, which encouraged them to turn to many Western foods. For example, there was a dramatic shift from eating sweet manioc at all meals, to eating new foods such as rice, Italian pasta, bread, oil, sugar, coffee, and canned meats.[29] The case of the Barí in relation to their environment, in terms of sustainability and conservation, illustrates a particular problem many indigenous people are likely to face in the future. The Barí people are one of the five thousand indigenous peoples (or nations) in the world who share this experience.[30] Because of the stereotype of the "Noble Savage," governmental officials have seen the Barí as natural conservationists who, if left alone, will voluntarily protect the biodiversity of Perijá National Park, where they live. At the same time, the stereotype causes denial about some of the ecological problems that they might be facing. The problem is that many indigenous people are facing changes in their population and environment as a result of outside influence. There is an urgent need to provide a more objective description of these societies and acknowledge the difficulties they face, in order to move forward in the search for solutions rather than disseminating romantic illusions.

The loss of forest in the Barí territory has been more dramatic than in many other indigenous societies. In the early 18th century, the Barí had a territory covering 25,000 square kilometers of rainforest (mostly lowland with some mountain evergreen forest).[31] From the 1790s to the 1980s, the Barí lost territory gradually. By the 1980s, they had lost 92 percent of their territory (23,100 square kilometers). Almost all of the lost territory was deforested for cattle pasture, except for a few patches along the river and inundated areas.[32] The remaining 8 percent of their territory (1,900 square kilometers) is mostly forested. In addition to the ongoing threat from cattle ranchers, there is the danger posed by logging companies (such as in the case described by Paul Cox[33] among the Samoans) and mining. It is clear that nonindigenous people are the major cause of destruction of the forest.[34] This has been documented previously.[35] However, the question here is what will happen to the rest of the Barí and their forest in the future?

There are certain prerequisites to a balance in the relationship between a human population and its natural environment. According to Leslie Sponsel, "[t]raditional indigenous populations are low in density and fairly mobile, and they practice a rotational subsistence economy with polycropping, adequate fallow periods, and ample areas of forest held in reserve for future gardens."[36] Most scholars agree with this statement. Slash-and-burn agriculturalists who practice hunting need a low population density—less than 0.2 people per square kilometer—in order to sustain their use of the forest. In the past century, the Barí population both had a low population density and practiced rotational agriculture. Unfortunately, the Barí population has been experiencing drastic changes in both areas. This is the essence of the human dilemma that most biologists and ecologists are facing in the world.[37]

The Barí population in the 18th century was estimated by Beckerman to be about 2,000–2,500 people.[38] Due to the loss of territory and massacres experienced from the 1770s to 1959, the Barí population was reduced to about 1,100 by 1960.[39] In addition, the introduction of western diseases, especially measles and influenza, killed about one quarter (or possibly more) of the Barí population in the early 1960s. Therefore, by 1964, the Barí population was at its lowest point, with 850 people. In 1982, their population increased to 1,560 people in Colombia and Venezuela.[40] By 1992, there were approximately 1,950 Barí.[41] Today, the estimated Barí population is approximately 3,000. Hence, although the Barí were at risk of vanishing in the 1960s, their population has recovered, increasing more than threefold in nearly thirty-four years.

With the change in the size of the Barí population, there has been a conspicuous modification in their relationship to resources available in their environment, the forest. The carrying capacity, which is the ability of a population to be sustained indefinitely in a specific territory, has been disrupted. Consider how the changes of Barí population size and territory have dramatically modified population density from the eighteenth century to the present. In 1700, the Barí population density was 0.1 person per square kilometer. This density is considered low and their population impact on their environment was minimal. By 1998, the Barí had a density of 1.2 persons per square kilometer, or twelve times what it had been in 1700.[42]

Let me illustrate the significance of this development with a hypothetical scenario involving the change in population density for a village of twenty-eight people in relation to one particular food resource, the tapir. If this hypothetical Barí village is harvesting two tapirs annually from a local population of twelve tapirs, they might not necessarily deplete the tapir population, since the tapir population might still be able to reproduce itself. If the Barí population increases by twelve times, from 28 to 336 people, they are likely to need twelve times more food. If they were to increase their tapir hunting in order to get the same amount of meat per capita, they would have to take twenty-four tapirs annually. Since there are only twelve tapirs in the local region, they would only have enough meat for the first six months, after which the tapirs would be gone.

The same would happen to other game and to fish, firewood, and many additional resources. In fact, this is happening to the Barí people. Michael Baksh describes the same problem among the Machiguenga of Peru, where a group was required to settle in a village permanently. After seven years, the village had increased to 250 people, even though "traditional settlements typically totaled 25 to 35 individuals."[43] Baksh also observes, "[t]he most apparent change in food production . . . in 1988 was a decline in wild food procurement. After having exploited the new site for seven years, people had nearly wiped out local fish and forest game."[44] William Vickers observed that the impact among the Siona-secoya (of Ecuador) of settling in an area for ten years was a serious depletion of major game.[45] In another example, Bernard

Nietschmann observed that among the Miskito Indians of Nicaragua "over-exploitation of land and sea fauna is just being felt . . . as cash alternatives are lacking and as subsistence skills are extended into market supply."[46] The same is happening to the large Barí villages: Saimadodyi, Bokshí, and Bachichida. This is unlikely to improve and rather will get worse with time; the Barí do not have other places to go, nor can they rely on a different subsistence strategy.

This leads to the following question: What will become of the Barí population by the middle of the next millennium? If the Barí population keeps increasing at the same pace as it did between 1982 to 1992, it will geometrically double every eighteen years, so that by the year 2050, it will be approximately 19,000. This would mean that the density would be 10 people per square kilometer, or 100 times higher than in the eighteenth century. This also means that the Barí either have to take drastic steps in changing their lifestyle or prevent this population increase. We cannot ignore the fact that the Barí population will likely continue to grow, since having children is very important in their culture. The impact on the rainforests will be serious, as it would mean a higher rate of resource use and depletion, besides a need to clear more land for gardens and to plant cash crops.

Changes in the rainforest and lifestyle will also affect the carrying capacity of the Barí population. These changes are all intrinsically involved in the process of deforestation, as a cause and effect of population changes. Even though the deforestation by cattle ranchers is glaringly obvious when one drives or flies through the region, we cannot ignore the fact that the Barí are also participating (by necessity) in deforesting their territory, albeit in a minor way. Why are they cutting down their forest when it provides them with the essential resources for their subsistence?[47] The Barí have started to become more sedentary in the past thirty years and have changed to a pattern of more extensive deforestation around their villages in order to plant their crops.

Besides the fact that they need more land for their gardens, there are three other main reasons for the Barí deforestation of their territory. The first reason is a product of the recommendations that both the Jesuit missionaries and the Venezuelan government have made to the Barí since the time of the first contact in 1960. The Jesuit missionaries and the Venezuelan government wanted the Barí to become more civilized and productive. They were perceived as uncivilized and unproductive, although they were self-sufficient horticulturalists and fishermen-hunters. Under the Jesuit model, in order to become integrated into the Venezuelan economy, the Barí had to produce cash crops and raise cattle. They were continually bombarded with these notions after contact and thus started to cut down the forest in the late 1960s. The loss of forest is quite obvious, especially around the villages that were founded as missions (Bokshí, Saimadodyi, and Bachichida). These villages are surrounded by extensive areas of depleted forest. For example, the average proportion of deforestation around small villages (20 to 50 people) is 24.7 percent, which is half what it is for the larger villages (120 to 320 people), 49.6 percent.[48]

The second reason for deforestation is the incentive to earn cash in order to pay for Western medicine, formal education for their children, and Western goods. This quest increased in the 1980s and 1990s. Exposed to an epidemic of Western diseases, their dependency on Western medicine increased vastly. Even though the Venezuelan government is supposed to supply medicine, the Barí have received insufficient amounts and often medicines they do not need (e.g., drugs to treat high blood pressure, which is a rare condition among them). The access to Western education is perceived by the Barí as a way for their children to increase their power to integrate into the Venezuelan society. Moreover, most of the highly educated Barí get better jobs working, for example, as school teachers or as government employees in the village. Thus, education is seen to enhance opportunities for increased income and for political power in their villages. Besides the need for formal education, the need for Western goods like steel instruments, flashlights, western clothing, store-bought food, shotguns, and shotgun shells is increasing. It is logical to assume that most Barí need more and more cash as they engage more in the Western market and gain tastes for Western goods. This was also observed by Baksh among the Machiguenga.

Moreover, with the increasing need to bring produce to the market, the Barí have the added need for more efficient forms of transportation. The Barí, living in villages that are several hours by foot from the roads, started buying mules to be able to carry their cash crops. This is becoming the standard mode of transportation today, even though mules cost the equivalent of six months income for a Barí working in a nearby ranch.

The Barí living six to seven hours upriver from the roads (approximately 22 percent of the population) need outboard motors, which are prohibitively expensive ($2,500 for a 45 horsepower motor), for their canoes, in order to bring their produce to the market. These motors are very difficult for a Barí to buy, since doing so means about five years of selling their labor at a cattle ranch. In order to achieve ownership of these expensive goods, the Barí have been looking for more efficient ways to earn more cash than they can by growing cash crops. For this reason, logging species of precious woods, specifically West Indian mahogany *(Swietenia mahagoni)* and Colombian mahogany *(Cariniana pyriformis),* has become a strategy for them, since mahogany is abundant in their territory and a commodity in the Venezuelan market. Moreover, to use and maintain mules and outboard motors, the Barí need to buy equipment and medicine for the mules and parts and gasoline for the outboard motors. To afford these, they need to sell cash crops such as plantain, banana, manioc, and pineapple. This increases the need to plant gardens and to cut down the forest.

The third reason for deforestation is to claim land as a personal property by establishing plantations and pasture for cattle grazing. This is a new trend among the young educated Barí. Claiming individual plots of land, called *fundos,* is a new development related to the notion of Western individualism among Barí, and many Barí people have started to acquire plots this way throughout their territory. Some villages still maintain their communal sense by maintaining group ownership of properties and working together in many communal activities (e.g., hunting, fishing, and keeping pasture for mules). However, it is obvious that the communal sense of property ownership is giving way to a more individualistic pattern. One factor in this shift is that the Barí are making sure their descendants have land for themselves. Even though some communities are proud to show a strong communal sense, many of their members have their *fundos.* And with the establishment of these *fundos,* more forest is cut down.

Moreover, the Barí are faced with an increasing arrival of Colombian and non-Barí indigenous peasants into the region in the past decade. These peasants are claiming the forested land next to or in Barí territory and are clearing it. Any peasant can claim the lands by simply "improving" it. By Venezuelan law, the improvement of vacant land implies deforesting it and converting it to cattle pasture or agricultural fields. For the Barí, this has been another major incentive to cut down the forest in their territory, in order to protect the land from being taken away.

In addition to invasions by peasants, the Barí are currently facing the beginning of coal mining exploitation in their territory. The Venezuelan government sold seven large coal concessions to four companies: Maicca, Cargoca, Corpozulia, and Consulminca. The open pit coal mining covers a large territory of 165,000 hectares, including a vast tract of forest and a dozen Barí villages and hamlets. In the maps that located the concessions, it was written that the areas were without human population. The simple act of allowing mining in a large sector of forest contradicts the notion of the pledge to protect the forest, made by the local environmental organization and the national governmental environmental agency (Ministerio de Recursos Naturales). Also, these concessions are where the largest tracts of forest are standing outside the National Park of Perijá, where a good portion of the Barí population lives or forages regularly.[49]

Many indigenous people are aware of the consequences of forest destruction and are taking steps toward finding ways to use the forest without destroying it.[50] The Barí are also aware of this pressing issue; the protection of their forests was one of the four items listed in a letter requesting all the Barí to gather in a meeting in July 1994. However, the Barí are experiencing a conflict between the twin goals of making a living and protecting their natural habitat. The Barí lived in an ecological context when the forest was unlimited and cutting it down was not a problem. They did not need to think about the notion of sustainability, because forest resources were always abundant. The notion of sustainable use is only developed when a society faces a scarcity of resources and takes measures for conserving them for continual use. The Barí ecological context has changed and the Barí perceive a decrease of rainforest resources. In some villages, people cannot even find resources for their basic traditional crafts like baskets and mats. Hunting game has become increasingly difficult, because game is becoming scarce in the forest, due to overhunting and shrinking habitat. The Barí have to travel up to ten kilometers in order to capture some game. Their priority is to feed their families, even though they are aware that their resources are quickly becoming depleted. Becoming completely sustainable is impossible because it would mean starvation. Nietschmann has observed the same conflict among the Miskito Indians in terms of overhunting and overfishing in Nicaragua. He states that "[s]ome Miskito are aware of the ecological blind alley they are entering by becoming dependent on declining resources. But they have few alternatives which provide monetary return."[51] The Barí have the same predicament.

Moreover, most of the territory that belongs to the Barí is an indigenous reserve as well as a national park (Parque Nacional Sierra de Perijá) covering an area of 295,000 hectares. According to the internal regulations imposed by the Institute of National Parks (IMPARQUES), the Barí are supposed to just practice "traditional" subsistence (e.g., hunting with bows and arrows instead of shotguns) and should not sell anything that they produce. Yet they are no longer a "traditional" people, but partially westernized. They have for decades been urged to become more integrated into the cash economy. One of the park rangers, a Barí man, told me that he needed to clear some forested land for cattle pasture because he could not make it with the salary he was receiving. For the same reason, Redford and Stearman state, "[v]irtually all [indigenous people] are now linked to the market economy through barter or actual cash exchange. [They] commonly want and have a right to health care, education, and material conveniences that improve their quality of life."[52] My question is whether we have the right to stop indigenous people from becoming westernized and to expect that they will protect the forest by using it in a sustainable way.

We have to be careful of the implications of portraying all indigenous peoples as sustainable societies. Indeed, indigenous societies are in many ways relatively sustainable when compared to our own societies in terms of pollution production and overconsumption of many resources. For example, for every two Americans there is one car,[53] compared to three hundred Barí per one car. This is just one example of the differences that make the Western lifestyle so enticing and make it very hard for Barí to choose to live a sustainable way of life as part of the conservation of biodiversity and culture.

CONCLUSION

Luisa Maffi[54] clearly states that indigenous people, anthropologists, and environmentalists know that the only way to protect these natural areas is for all parties to open a dialogue to search for the best alternative without sacrificing their desires. This is happening in many places, where NGOs and governments are working with indigenous people to find constructive alternatives.[55] But these dialogues have not taken into account that the ecological conditions of indigenous people are changing quite rapidly.

As previously stated, there is a major contradiction between the lifestyles of indigenous people required for a sustainable relationship to their forest and our current Western lifestyle. We should be aware that people in the developed nations cannot continue living this irresponsible way of life. It is clear that the entire world population cannot achieve the level of consumerism currently enjoyed by the developed nations. The resources needed for this are simply not available.[56] Shridath Ramphal and Steven W. Sinding argue that developed nations consume twenty to thirty times more resources per capita than "residents of the poor, developing nations."[57] In fact, the gap is even larger when you contrast people of developed nations and indigenous people.

Redford and Stearman also observe that "[t]o expect indigenous people to retain traditional, low-impact patterns of resource use is to deny them the right to grow and change in ways compatible with the rest of humanity."[58] In addition, Ramphal and Sinding state, "from a consumption standpoint, at present levels of consumption, . . . 5 percent [of the future growth in the developed world] will impose on the planet a greater burden than the 95 percent born in the developing world." Here lies an unbalanced relationship of a small population consuming a larger amount of resources while another small population is being held responsible for protecting the biodiversity of the majority of the world's resources by living in a sustainable way. This is the contradictory essence of green imperialism.

We expect indigenous people to carry the burden of forest conservation, portraying them as "Noble Savages," while people in developed nations lead irresponsible lives, consuming three quarters of the world's resources. "How much more environmental abuse can ecosystems and societies tolerate? Western societies have long demanded that indigenous societies change; the time is long overdue for indigenous societies to demand some changes in Western societies."[59] Therefore, conservation of natural ecoregions will not be possible in the land of indigenous people or developing nations unless access to resources and commodities is made more equal on the planet. In order to encourage conservation of natural areas and make this possible, developed nations need to reduce their consumption of resources and help both developing nations and indigenous people to improve their living standards.

52   Chapter 3

NOTES

Special thanks to Gerald Visgilio, Diana Whitelaw, and Glenn Dreyer for inviting me to write a chapter for this volume. I would like to thank all the Barí people for letting me reside and conduct my research in their cultural and natural environment—with special thanks to Andrés Achirabu, Jaime, the late Luisa Oshkoro, Adrián Okboo, the late Nora, Akirida, and David Aleobadda. I am quite grateful to my wife, Anne-Marie Lizarralde, for her support and the meticulous editorial assistance with this chapter. Any remaining errors in the text are my sole responsibility.

1. For more details, see Lester R. Brown, "The Acceleration of History," in *State of the World,* ed. Lester R. Brown, 3–20 (New York: Norton, 1996), and Clive Ponting, *A Green History of the World: The Environment and the Collapse of Great Civilizations* (New York: Penguin, 1991).

2. For more information on indigenous peoples' relations to the environment, see Gonzalo Oviedo, Luisa Maffi, and Manuel Lizarralde, project coordinators, "Indigenous and Traditional Peoples in the Global 200 Ecoregions," a 200 cm by 120 cm world map (Washington: World Wildlife Fund, 2000), and Luisa Maffi, ed., *On Biocultural Diversity: Linking Languages, Knowledge and the Environments* (Washington, D.C.: Smithsonian Institution Press, 2001).

3. For more details, see William Cronon, *Changes in the Land: Indians, Colonists, and the Ecology of New England* (New York: Hill and Wang, 1983); Shepard Krech, *The Ecological Indian: Myth and History* (New York: Norton, 1999); Maffi, ed., *On Biocultural Diversity;* Eglée Lopez Zent and Stanford Zent, "Amazonian Indians as Ecological Disturbance Agents: The Hoti of Venezuela," in *Ethnobotany and Conservation of Biocultural Diversity,* ed. Thomas Carlson and Luisa Maffi (New York: New York Botanical Garden, forthcoming).

4. A number of works illustrate the sustainable relationship of indigenous societies to their environment: Eugene N. Anderson, *Ecologies of the Heart: Emotion, Belief, and the Environment* (New York: Oxford University Press, 1996); Jason W. Clay, *Indigenous Peoples and Tropical Forests: Models of Land Use and Management from Latin America* (Cambridge, Mass.: Cultural Survival, 1988); Suzanne Head and Robert Heinzman, eds. *Lessons of the Rainforest* (San Francisco: Sierra Club Books, 1990); Elizabeth Kemf, ed., *The Law of the Mother: Protecting Indigenous Peoples in Protected Areas* (San Francisco: Sierra Club Books, 1993); Lopez Zent and Zent, "Amazonian Indians"; Gonzalo Oviedo, "Los Pueblos Indígenas y la Conservación: Declaración de Principios del WWF" (Gland, Switzerland: World Wildlife Fund), 1996; Darrell A. Posey, *Biological and Cultural Diversity: The inextricable, Linked by Language and Politics* (Oxford: Oxford Centre for the Environment, Ethics and Society, Mansfield College, 1996); Kent H. Redford and Christine Padoch, eds., *Conservation in Neotropical Forests: Working from Traditional Resources Use* (New York: Columbia University Press, 1992); Leslie E. Sponsel, ed., *Indigenous Peoples and the Future of Amazonia: An Ecological Anthropology of an Endangered World* (Tucson: University of Arizona Press, 1995); Victor Manuel Toledo, "What Is Ethnoecology? Origins, Scope, and Implications of a Rising Discipline," *Etnoecología* 1, no. 1 (1992): 5–21.

5. Stan Stevens, ed., *Conservation through Cultural Survival: Indigenous Peoples and Protected Areas* (Washington, D.C.: Island Press, 1997), 2.

6. Marianne Schmink, Kent H. Redford, and Christine Padoch, "Traditional Peoples and the Biosphere: Framing the Issues and Defining the Terms," in *Conservation in Neotropical Forests*, ed. Redford and Padoch, 3–34.

7. Some authors show that the noble savage notion is not homogeneously represented by indigenous societies: Stephen Beckerman, "On Native American Conservation and the Tragedy of the Commons," *Current Anthropology* 37, no. 4 (1996): 659–61; John H. Bodley, ed. *Tribal Peoples and Development Issues: A Global Overview*, 3d ed. (Mountain View, Calif.: Mayfield, 1988); Katharine Milton, "Civilization and Its Discontents: Amazonian Indians Experience the Thin Wedge of Materialism," *Natural History* 3 (1992): 36–45; Manuel Lizarralde, "500 Years of Invasion: Eco-Colonialism in Indigenous Venezuela," *Kroeber Anthropological Society Papers,* 75–76 (1992): 62–79; Manuel Lizarralde, "Ethnoecology of Monkeys among the Barí of Venezuela: Perception, Use and Conservation," in *Primates Face to Face: The Conservation Implications of Human and Nonhuman Primate Interactions,* ed. Agustin Fuentes and Linda D. Wolfe, 85–100 (Cambridge: Cambridge University Press, 2002); Schmink et al., *Conservation in Neotropical Forests;* and Leslie E. Sponsel, "Myths of Ecology and Ecology of Myths: Were Indigenes Noble Conservationists or Savage Destroyers of Nature?" (paper presented at the Second Annual Conference on Issues of Culture and Communication in the Asia/Pacific Region, University of Hawaii 1992).

8. Krech, *Ecological Indian,* 228.

9. Krech, *Ecological Indian.*

10. Dennis Martinez, "First People, Firsthand Knowledge," *Sierra* 81, no. 6 (1996): 50–51, 70–71, quoted in Krech, *Ecological Indian,* 228.

11. Martinez, "First People," 50–51, 70–71, quoted in Krech, *Ecological Indian,* 228.

12. Krech, *Ecological Indian,* 228.

13. The rate of destruction of the world rain forest in 1989 was approximately 1.8 percent. Edward O. Wilson, *The Diversity of Life* (New York: Norton, 1992), 275; Norman Myers, *The Primary Source: Tropical Forests and Our Future* (New York: Norton, 1992), xvii–xviii, quoted in Leslie E. Sponsel, Robert C. Bailey, and Thomas N. Headland, "Anthropological Perspectives on the Causes, Consequences, and Solutions of Deforestation," in *Tropical Deforestation: The Human Dimension*, ed. Leslie E. Sponsel, Thomas N. Headland, and Robert C. Bailey, 3–52 (New York: Columbia University Press, 1996), 3. Each year, we lose 142,200 square kilometers of forest. Approximately 54 percent of the prehistoric world forest has been destroyed. Most of the deforestation (85 percent in the Brazilian Amazon) is done by logging companies, mining projects, and cattle ranchers (Susanna Hecht, "The Sacred Cow in the Green Hell," *Ecologist* 19, no. 6 [1989]; Susanna Hecht and Alexander Cockburn, *The Fate of the Forest: Developers, Destroyers and Defenders of the Amazon* [New York: Harper Perennial, 1990], quoted in Sponsel, *Indigenous Peoples,* 265). When we look at particular regions, like the Barí territory, it is ever more serious, since 92 percent of their original eighteenth-century territory is deforested today.

14. Cf. Manuel Lizarralde, "Biodiversity and the Loss of Indigenous Languages and Knowledge in South America," in *On Biocultural Diversity,* ed. Maffi, 265–81; and Lizarralde, "Ethnoecology of Monkeys among the Barí of Venezuela," in *Primates Face to Face,* ed. Fuentes and Wolfe, 85–100.

15. This is clearly stated in the work of Brown, ed., *State of the World,* and Ponting, *Green History.* For example, Brown states that the U.S. standard of living is

54   Chapter 3

ecologically possible for only two billion people on the earth. There are not enough resources, especially energy and water, to support a larger population with the equivalent lifestyle. However, many people from developing nations would like to move to North America or Europe to gain access to their commodities and lifestyle; the misery experienced by these people is not just a choice but a socioeconomic product of the historical process of colonization started in the sixteenth century. Similarly, indigenous people want to become westernized and have access to Western technology and commodities to make their lives easier.

16. Marshall Sahlins, "Notes on the Original Affluent Society," in *Man the Hunter,* ed. Richard Lee and Irven DeVore, 85–89 (New York: Aldine, 1968).

17. In the 1960s, anthropologists made popular the notion that "primitive" societies had an easy way of life, having to work only three to four hours a day (see Sahlins, "Notes"). It is true that members of indigenous societies sometimes lead such a way of life, but only when resources are abundant and accessible. Normally, the mortality is high, as young children and elders die at higher proportions than in our society, keeping the population low naturally. The notion of the affluent society was overused and grossly overgeneralized in the literature, so much that many readers and other disciplines took it as the rule for all indigenous societies.

18. John H. Bodley, *Victims of Progress,* 3d ed. (Mountain View, Calif.: Mayfield, 1990), 155.

19. This is clearly stated in Bodley, ed., *Tribal Peoples;* Brian Furze, Terry De Lacy, and Jim Birchhead, *Culture, Conservation and Biodiversity: The Social Dimension of Linking Local Level Development and Conservation through Protected Areas* (New York: Wiley, 1996); Redford and Padoch, eds., *Conservation in Neotropical Forests;* and Stevens, *Conservation.*

20. Lopez Zent and Zent, "Amazonian Indians," presents the best documented example of sustainable management of a forest by those traditionally foraging there.

21. Lopez Zent and Zent. "Amazonian Indians."

22. Lizarralde, *Biodiversity.*

23. This discrepancy in consumption is clearly documented in Brown, ed., *State of the World;* Ponting, *Green History;* and Vandana Shiva et al., *Biodiversity: Social and Ecological Perspectives* (London: Zed, 1991).

24. See Coordinadora de las Organizaciones Indígenas de la Cuenca Amazónica (COICA) "Two Agendas on Amazon Development," *Cultural Survival Quarterly* 13, no. 4 (1989): 75–87; Darrell A. Posey, "Interpreting and Applying the 'Reality' of Indigenous Concepts: What Is Necessary to Learn from the Natives?," in *Conservation in Neotropical Forests,* ed. Redford and Padoch, 21–34; Kent H. Redford and Allyn M. Stearman, "Forest-Dwelling Native Amazonians and the Conservation of Biodiversity: Interests in Common or in Collision?" *Conservation Biology* 7, no. 2 (1993): 248–55.

25. Bodley, *Tribal Peoples,* 6.

26. Many examples exist such as Paul Alan Cox and Thomas Elmqvist, "Ecocolonialism and Indigenous Knowledge Systems: Village Controlled Rainforest Preserves in Samoa," *Pacific Conservation Biology* 1 (1993): 6–13; Raymond Dasmann, "National Parks, Nature Conservation, and "Future Primitive," in *Tribal Peoples,* ed. Bodley, 301–10; Head and Heinzman, *Lessons;* Barbara Rose Johnston, ed., *Who Pays the Price?: The Sociocultural Context of Environmental Crisis* (Washington, D.C.: Island Press, 1994); Bonnie J. McCay and James M. Acheson, eds., *The Question of the Com-*

mons: The Culture and Ecology of Communal Resources (Tucson: University of Arizona Press, 1987); Susan Milius, "When Worlds Collide: Why Can't Conservation Scientists and Indigenous Peoples Just Get Along?" Science News 154, no. 6 (1998): 92–94; M. L. Oldfield and J. B. Alcorn, eds. Biodiversity: Culture, Conservation and Ecodevelopment (Boulder: Westview, 1991); Nancy Lee Peluso, Rich Forest, Poor People: Resource Control Resistance in Java (Berkeley: University of California Press, 1992); Redford and Stearman, "Forest-Dwelling Native Amazonians and the Conservation of Biodiversity," 248–55; Richards, The Tropical Rainforest, 1996; Shiva et al., Biodiversity; Sponsel et al. Tropical Deforestation; John Terborgh, Requiem for Nature (Washington, D.C.: Island Press, 1999); Jorge Ventocilla, Heraclio Herrera, and Valerio Núñez, Plants and Animals in the Life of the Kuna (Austin: University of Texas Press, 1995).

27. Clay, Indigenous Peoples and Tropical Forest; Jason W. Clay, "How Reserves Can Work," Garden 13, no. 5 (1989): 2–4; Clay, "Indigenous Peoples: The Miner's Canary for the Twentieth Century," In Lessons of the Rainforest, ed. Suzanne Head and Robert Heinzman, 106–17 (San Francisco: Sierra Club Books, 1990); and Clay, "Resource Wars: Nation and State Conflicts of the Twentieth Century," in Who Pays the Price? ed. Johnston, 19–30.

28. The main problem is that the current belief that indigenous people are the stewards of world biodiversity assumes that they will not change their consumption levels and will preserve the biodiversity for us at no cost. The reality is that indigenous peoples possess neither the sociopolitical nor ecological status to carry this responsibility. This view is complemented by the following quote: "[a]s many Kuna [an indigenous people of Panama] see it, the problem comes in part through loss of knowledge, the intimate knowledge of forest and marine environments retained by the oldest hunters, fishermen agriculturists, and especially, medicinal curers. As many young Kuna abandon traditional subsistence for paid labor, either in Kuna Yala or in the city, and as almost no one under thirty seriously apprentices to learn traditional medicine, the store of environmental knowledge rapidly diminishes and, with it, the intimacy of the Kuna's connection with the natural world (Ventocilla et al., Plants and Animals, x). This is better illustrated by the Barí case in the following section of this chapter.

29. The source for this information is my twenty-eight months of fieldwork starting in 1988, as well as my exposure to my father, Roberto Lizarralde, who did the first contact of the Barí in 1960 and performed forty-two years of fieldwork. I have been with the Barí since childhood, when I assisted my father in his fieldwork. Also, I have had long discussions with Stephen Beckerman, who has also done extensive fieldwork with the Barí. I am in a fortunate situation to have access to an extensive source of information not published or published in many obscure places.

30. Cf. Clay, Resource Wars, 24.

31. Stephen Beckerman, "Datos Etnohistóricos Acerca de los Barí Motilones," Colección de Lenguas Indigenas, Serie Menor, vol. 7 (Caracas: Montalbán, 1978), 52.

32. Roberto Lizarralde, "Barí Settlement Patterns," Human Ecology 19, no. 4 (1991): 428–52, and Roberto Lizarralde and Stephen Beckerman, "Historia Contemporánea de los Barí" Antropológica 58 (1982): 3–51.

33. Paul A. Cox, Nafanua: Saving the Samoan Rain Forest (New York: Freeman, 1997).

34. Four cases observed in the field illustrate this destruction. One Colombian peasant destroyed more than two hundred hectares of forest in the early 1990s in the Serrania of Abusanqui, fifteen kilometers north of Rio de Oro. Another Colombian peasant cut down four hundred hectares mid-1990s, seventeen kilometers south of the Ariquaisa River and the village of Karanyakaig. A couple of cattle ranchers destroyed twenty-four thousand hectares of forest between the Ariquaisa River and Santa Rosa and on the east slope of Serrania Abusanqui, in 1999 and 2000. The last case occurred near the Ariquaisa River and south of the Kumagda village, where a cattle rancher has destroyed two hundred to three hundred hectares of forest between 1998 and 1999 and continues to bulldoze more forest in the present. All these cases occurred inside the Barí territory.

35. M. Lizarralde, "The Barí Responses to Mineral and Lumber Concessions in Venezuela," in *El Dorado Revisited: Gold, Oil, Environment, People and Rights in the Amazon,* ed. Leslie Sponsel (Washington, D.C.: Island Press, forthcoming in 2003), and R. Lizarralde. "Barí Settlement."

36. Sponsel, *Indigenous Peoples,* 265.

37. Marjorie L. Reaka-Kudla, Don E. Wilson, and Edward O. Wilson, eds. *Biodiversity II: Understanding and Protecting Our Biological Resources* (Washington, D.C.: Joseph Henry, 1997); Paul W. Richards, *The Tropical Rain Forest: An Ecological Study,* 2d ed. (Cambridge: Cambridge University Press, 1996); Terborgh, *Requiem for Nature;* Wilson, *Diversity of Life.*

38. Beckerman, *Datos Etnohistóricos,* 51.

39. R. Lizarralde and Beckerman, "Historia Contemporánea," 49.

40. R. Lizarralde and Beckerman, "Historia Contemporánea," 49.

41. Cf. R. Lizarralde, "Barí Settlement," 437.

42. Cf. Beckerman, *Datos Etnohistóricos,* 1978, Lizarralde and Beckerman, "Historia Contemporánea."

43. Michael Baksh, "Change in Machiguenga Quality of Life," in *Indigenous Peoples and the Future of Amazonia: An Ecological Anthropology of an Endangered World,* ed. Leslie E. Sponsel, 187–205 (Tucson: University of Arizona Press, 1995), 197.

44. Baksh, "Change," 197.

45. William T. Vickers, "Hunting Yields and Game Composition over Ten Years in an Amazon Indian Territory," in *Neotropical Wildlife Use and Conservation,* ed. J. G. Robinson and K. H. Redford, 53–81 (Chicago: University of Chicago Press, 1991).

46. Bernard Nietschmann, *Between Land and Water: The Subsistence Ecology of the Miskito Indians, Eastern Nicaragua* (New York: Seminar Press, 1973), 180.

47. It is known that slash-and-burn horticulturalists need to clear their fields in order to produce starchy food. The practice of slash-and-burn horticulture does minimum damage to the forest, since it uses small patches of land that are sparsely distributed. After the field is abandoned, the nearby forest will quickly reconquer this clearing (Lopez Zent and Zent, "Amazonian Indians").

48. This figure was adapted and extrapolated from Clifford A. Behrens, Michael G. Baksh, and Michel Mothes, "A Regional Analysis of Barí Land Use Intensification and Its Impact on Landscape Heterogeneity," *Human Ecology* 22, no. 3 (1994): 303.

49. M. Lizarralde, "Barí Responses," and M. Lizarralde, *Ethnoecology.*

50. There are many example of forest protection or concern about it: COICA, "Two Agendas"; Richard K. Reed, *Forest Dwellers, Forest Protectors: Indigenous Models for International Development* (Boston: Allyn and Bacon, 1997); Redford and Stearman, "Forest-Dwelling Native Amazonians," Sponsel et al., *Tropical Deforestation;* and Ventocilla et al., *Plants and Animals.* For example, The Kuna believe "[t]he earth is the mother of all things. . . . [S]ocial change [needs to be viewed] as a means of saving traditional ecological knowledge and 'returning' it to the community. . . . While the Kuna have a tradition of living in harmony with the land, the intrusion of the market economy is eroding the very basis of their sustainable way of life. As a response to this crisis, th[ese authors] seeks to develop native self-awareness and provide a model for collaboration between indigenous peoples and foreign researchers" (Ventocilla et al., *Plants and Animals,* back cover).

51. Bernard Nietschmann, "Hunting and Fishing Focus among the Miskito Indians, Eastern Nicaragua," *Human Ecology* 1 (1972): 66.

52. Redford and Stearman, "Forest-Dwelling Native Amazonians," 252.

53. Ponting, *Green History,* 340.

54. Maffi, ed., *On Biocultural Diversity.*

55. Katrina Brandon, Kent H. Redford, and Steven E. Sanderson, eds., *Parks in Peril: People, Politics and Protected Areas* (Washington, D.C.: Island Press and the Nature Conservancy, 1998); Furze et al., *Culture, Conservation and Biodiversity;* Stevens, *Conservation.*

56. For more details, see Brown, *State of the World;* Ponting, *Green History;* Shridath Ramphal and Steven W. Sinding, "Conclusions," in *Population Growth and Environmental Issues,* ed. Shridath Ramphal and Steven W. Sinding (Westport, Conn.: Praeger, 1996).

57. Ramphal and Sinding, *Population Growth,* 173.

58. Redford and Stearman, "Forest-Dwelling Native Amazonians," 252.

59. Sponsel, "Myths of Ecology," 32.

II

EMPIRICAL RESEARCH AND
METHODOLOGICAL CHALLENGES

4

Burning and Burying in Connecticut: Are Regional Solutions to Solid Waste Disposal Equitable?

Timothy Black and John A. Stewart

To "think globally" and "act locally" has become a rallying cry for the environmental movement in the United States. Dominated largely by the white middle class, the environmental movement has nonetheless begun to pick up steam in racial minority communities, where charges of environmental racism have been leveled at economic and political elites. International environmentalism has drawn attention to depleted world resources, disappearing rainforests, global warming, and neglectful fishing, while minority groups have embraced environmental justice to protest the siting of hazardous and solid waste disposal facilities, energy generators, sewage treatment and chemical plants in or near their neighborhoods. These latter concerns received support from President Clinton in 1994, when he ordered federal agencies to reduce environmental injustices that have disparate impact on racial minority communities. Subsequent Environmental Protection Agency (EPA) guidelines have empowered its Civil Rights division to decide on discrimination cases filed under Title VI of the Civil Rights Act of 1964. Despite opposition from state leaders, who argue the policy has resulted in a proliferation of court cases and has stymied commercial development in poor urban areas, then–Vice President Gore reinvigorated the administration's directives on Earth Day in 1998, arguing that "there have been strong expressions of concern from community leaders that our efforts to date have not been sufficient."[1]

Researchers have taken up these concerns and attempted to empirically assess the claims of environmental racism or environmental inequity. This research, still in its infancy, has generated a variety of methodological strategies and debates about how to assess environmental justice. No one strategy is likely to emerge as the best methodology; together, however, these strategies can help to inform the policy debates concerning the production, storage, transfer, and disposal of waste. These debates are essential to understanding unequal distributions of burden, just as they are useful in the development of interpretive frames for understanding the interrelationship between economic production, land use, and population distribution.

Our work in Connecticut focuses on the rapidly changing system for disposing of municipal solid waste. The old town dumps are obsolete and being replaced with a regional system for processing waste. New technologies, federal legislation, and declining landfill area are key forces driving these changes. State agencies and waste facility operators champion these new developments as environmentally safer, while neighborhood groups protest the use of their communities as dumping grounds for waste generated in other towns, even other states. In this chapter, we document these changes by identifying the location of all facilities in Connecticut that constitute its system of solid waste management and the distribution of populations around these facilities. Further, we particularly focus on newer developed regional facilities to determine if these facilities are being disproportionately placed in minority or lower income communities. Connecticut provides a unique opportunity for research to inform public policy on these issues, because the selection and development of regional facilities are continuing today. More generally, environmental justice studies that focus on the state level are particularly relevant to policy analysis, because state agencies make or state regulations constrain many of the decisions about the location of waste facilities. If environmental equity is a serious concern among state agencies and the public, then it is vital that independent researchers carefully document site locations and test for disproportionate burden.

ENVIRONMENTAL EQUITY STUDIES

While several environmental equity studies, especially concerning air pollution, were done in the '70s, the first study to receive widespread recognition was conducted by the United Church of Christ (UCC) in 1987. The UCC study was the first national study to assess the location of hazardous waste facilities. This groundbreaking study supported the claim of environmental racism by establishing that minorities were disproportionately represented in zip code areas containing commercial treatment, storage, or disposal facilities (TSDFs) for hazardous waste. The percentage of racial minorities in these zip code areas was twice as large as in areas without a facility and it was more than three times larger in areas where two or more facilities were located or where one of the five largest landfills was located. Moreover, a discriminant function analysis indicated that race was the best predictor of the presence of these facilities, even when including measures of social class.[2]

Douglas Anderton and his colleagues at the University of Massachusetts in Amherst followed with the next national study, but chose census tracts as their units of analysis instead of zip code areas. Census tracts are smaller geographical units that permit a more refined analysis of residential areas located in close proximity to waste facilities. Anderton et al. argued that this reduced the potential for "aggregation errors" or "ecological fallacies."[3] Their results differed from the UCC report. They compared the sociodemographic characteristics of 1980 census tracts in standard metropolitan statistical areas (SMSAs) containing commercial hazardous waste facilities to tracts without facilities in the same SMSA. They found that racial minorities were not disproportionately located in tracts with facilities and that minority percentages did not increase significantly in abutting tracts. Only when they included tracts within a radius of 2.5 miles of a facility were black and Hispanic populations significantly higher than in the rest of the SMSA.[4] Instead, the population living in tracts with commercial hazardous waste facilities disproportionately included fewer employed men, more adults working in manufacturing jobs, and a lower-valued housing stock among owner-occupied homes. When they limited their analysis to the largest twenty-five SMSAs, the same variables were significant, but they also found that blacks were significantly *less* likely to live in tracts with facilities, while Hispanics were *more* likely to reside in host tracts. Further, when they conducted the same analysis using 1990 census data, their study again indicated that more residents in host tracts were employed in manufacturing jobs and housing values were significantly lower, but in 1990 the percentage of families below the poverty line and the percentage of families residing in public housing were significantly higher in tracts with facilities. The factor of race, however, was not found to be significantly different.[5]

In each of their above analyses, Anderton and his colleagues at the University of Massachusetts found that the variable indicating the greatest difference between tracts with and without facilities was the percentage of employees in manufacturing employment. This was confirmed in their multivariate analysis. Using logit regression on their 1980 data, they found that the only variable that increased the odds of a tract containing a TSDF was an increase in the percentage employed in manufacturing employment. They concluded that racial minorities were not likely to live in neighborhoods with commercial hazardous waste facilities, but were more likely to be concentrated in areas one to three miles from these facilities. Instead, neighborhoods with facilities were more likely to be white, industrial, working-class neighborhoods that by the '90s were becoming poorer. Their analyses within EPA regions also largely supported their interpretation. In 1980, race was not found to be a predictor of facility location in any of the ten EPA regions in the country, while the percentage of manufacturing workers increased the likelihood of a tract containing a waste facility in seven of the ten regions. In their

1990 analysis, they did find significantly more blacks in host facilities in the EPA's southeast region and significantly more Hispanics in neighborhoods with facilities in the southwest. Nevertheless, employment in industrial or precision manufacturing occupations was significantly greater in host tracts in all ten of the EPA regions in 1990.[6]

The different outcomes reported by the UCC and U. Mass. studies set the stage for a plethora of studies that have attempted to distinguish racial and class effects on waste-facility sitings. Evan Ringquist used zip code areas in his national study of toxic release inventory (TRI) sources. In addition to several race and income variables, he included important control variables to predict three aspects of the TRI data: the presence of a TRI source, the number of TRI sources, and the amount of TRI releases. Furthermore, he compared the effects of using a national or state reference group in his analyses. The results strongly indicated that even after controlling for other variables, zip code areas with more African American, Hispanic, or poor residents were more likely to contain a TRI source or have larger chemical releases.[7] Unfortunately, he did not report standardized coefficients or actual significance levels, so it is not possible to assess the relative importance of these different predictors.

John Hird and Michael Reese conducted a national study of counties and their exposures to a large and diverse set of pollution indicators—twenty-nine in all—including the number and capacity of commercial landfills. Their multivariate analyses of each pollution indicator found a very consistent tendency for counties with more minorities to have higher pollution potentials, although other control variables, such as the percent of owner-occupied housing or the population density, were usually more important predictors.[8] In addition to these national studies, several regional studies of air pollution, Superfund sites, medical waste facilities, and landfills have also attempted to assess the relative effects of race and income on site locations,[9] but only a few studies have been conducted that examine population distribution around municipal solid waste disposal facilities. These facilities differ from commercial hazardous waste facilities in that the waste processed and disposed of cannot exceed specified quantities of hazardous waste. Contractors usually monitor this themselves with the expectation that they will not exceed regulatory guidelines. Siting these types of facilities in neighborhoods often provokes community resistance, or as Michael Greenberg explains, trash-to-energy plants "should be the type of pariah land use that so outrages most people that they would be sited into neighborhoods and towns occupied by relatively powerless people."[10]

Robert Bullard examined the siting of solid waste facilities in Houston, Texas. He found that permitted facilities were disproportionately located in predominantly racial minority neighborhoods and near schools with large racial minority enrollments.[11] Vicki Been provided a secondary analysis of

Bullard's work, claiming that he had mistakenly counted some facilities more than once in his analysis and that he had used "neighborhoods" as his unit of analysis without explaining how they were defined. Further, Been's study was particularly concerned with the issue of whether the siting process was guided by racial prejudice or by market dynamics.[12] Been estimated that 17 to 20 percent of U.S. households move each year and that it was therefore difficult to determine from cross-sectional data whether facilities were being sited in black communities or whether white flight combined with decreasing housing values in areas around facilities were producing higher concentrations of black residents in these areas. Been explained, "As long as the market discriminates on the basis of race, it would be remarkable if LULUs [locally undesirable land uses] did not eventually impose a disproportionate burden upon people of color."[13]

In her secondary analysis of Bullard's work, Been used census tracts as her unit of analysis and compared racial compositions in ten tracts with facilities (three incinerators and seven landfills) to the racial population in the city as a whole. She found that one half were sited in neighborhoods with significantly larger populations of African Americans than reside in the larger city. Further, she found that in three of the ten facilities, the poverty rates for the census tract were significantly greater than the overall poverty rate in the corresponding county. She also found that the percentage of African Americans residing in host tracts increased dramatically (223 percent) between 1970 and 1980 compared to increases in the overall African American population in the city (7 percent). Increases in the African American population in host tracts also continued in the 1980s, while the city population of African Americans stayed about the same. Been also demonstrated that by the 1990 census, seven of the host tracts had become significantly poorer and that median income had fallen more in nine of the ten host tracts when compared to the county population. While there is some support that the siting process itself had a disproportionate effect upon minority communities, there is also strong support that market dynamics that resulted in white flight and lower housing values also had a major effect after the facility was sited. Thus, Been underscores the importance of longitudinal studies that account for population changes that occur after the siting of facilities.

In addition to Bullard's research in Texas, there have been three national studies of solid waste facilities. Greenberg's study examined towns where trash-to-energy facilities are located and used both service areas outside of these towns and the U.S. population as comparisons. His article illustrates five criteria for measuring environmental equity. In applying his criteria, Greenberg found that the poor and minorities were more likely to reside in towns with large trash-to-energy plants than in the surrounding service area, but he found no differences for towns with smaller plants. When comparing all towns with facilities to the service areas and to the U.S. population, he found that per capita income was lower in towns with facilities across both comparisons. However, the results for minorities differed. Minority populations in towns with facilities were significantly greater than minority populations in the service areas, but lower than the minority population in the United States. However, when the population in towns was weighted in the analysis, the percent of minorities in towns with facilities was significantly greater than the minority population in the larger United States. To illustrate the importance of testing for different populations, he also found that the percent elderly was a better predictor of facility location than were race or poverty.[14]

In 1995, the U.S. General Accounting Office (GAO) published their study of nonhazardous waste facilities. Using census block groups—even smaller geographical areas than census tracts—as their unit of analysis, they analyzed block groups within one mile and then within three miles of 295 nonhazardous landfills throughout the nation. They found that racial minorities were less likely to reside within one mile or within three miles of landfills when compared to the racial minority population residing in metropolitan counties in the United States. In a second analysis, the GAO compared racial minority populations in block groups within one mile to the population in the remainder of the corresponding county. They found that the racial minority population within one mile of a landfill was greater than that of the surrounding county in only 27 percent of cases, while the median income was lower in 44 percent of cases. They concluded that areas around nonhazardous landfills throughout the nation were more likely to have fewer racial minorities and higher incomes compared to the surrounding county population.[15]

The most recent equity study of municipal solid waste disposal, including sewage treatment plants, was done by William Markham and Eric Ruffa.[16] They examined forty-nine randomly selected U.S. cities, comparing race, ethnicity, income, and housing variables in census tracts where facilities existed to town data. They characterized their study as an assessment of "waste streams," or as an examination of waste originators and waste recipients, in an effort to determine "whether the more privileged dumped their wastes on the less privileged."[17] In their assessment of landfills and incinerators, the results were in the opposite direction than they had expected, with fewer minorities and less poverty found in tracts with facilities. They did find a modest bivariate relationship between education and the destination tracts. This pattern held up in their examination of sewage treatment plants and in their separate analysis of facilities in the southern United States, with the exception that they did find Hispanics were slightly more likely to live in tracts with sewage treatment facilities. The most consistent finding across these analyses was that less-educated residents were more likely to live in facility tracts.

This brief review of previous studies indicates that research results are influenced by the types of waste facilities studied, the size—from local to national—of the region studied, and the methodologies used. Our study draws upon this research, especially in informing our methodology. We focus on one state, but our study is particularly timely, as we attempt to examine the state's transition to a regional system of solid waste disposal, which has occurred rapidly in the past fifteen years. Although the regional authority overseeing the majority of solid waste disposal was legislatively established in 1972, the first regional trash-to-energy plants and ash landfills did not begin operation until the mid-1980s. The rapidity and scope of this change is reflected by the closures of municipal solid waste (MSW) landfills during this time. In 1980, there were more than one hundred operating MSW landfills; in 1996, when we established our data set for this study, there were three remaining MSW landfills and six ash landfills. Similarly, our data set includes seventy-three bulky waste landfills, where waste and debris from land clearing activities are disposed, but the recent solid waste management plan published by Connecticut's Department of Environmental Protection indicates that the number of these facilities will likely be reduced to four in the near future. Thus, we are attempting to document ground that is constantly shifting as we collect data and write. While this can make research difficult, conducting research at a time in which so many facility sitings are occurring provides an opportunity to study the dynamics of social change. The research reported here is one snapshot of a rapidly changing system.

ANALYZING SOLID WASTE DISPOSAL IN CONNECTICUT

In our analysis, we include facilities in Connecticut that are used in the management of municipal solid waste. The different types of facilities are as follows:

1. RRFs: resource recovery facilities (also referred to as trash-to-energy plants) that incinerate solid waste and generate electricity
2. Ash LFs: ash landfills, where the ash produced from the incineration process is deposited
3. TSs: transfer stations, where waste is collected for transportation to RRFs
4. VRFs: volume reduction facilities, where construction and/or demolition waste is processed to recover materials that can be reused, recycled, or burned
5. IPCs: intermediate processing centers, where recyclable waste is gathered and bundled to be sold on the market
6. BWLFs: bulky waste landfills where construction and demolition debris are deposited

In this chapter, we examine whether environmental justice issues are apparent in what we refer to as the regionalization process for disposing of Connecticut's solid waste. We provide three different analyses. First, we examine the locations of seven RRFs along with their ash landfills, which together constitute the heart of the regional system. The second grouping combines the same facilities with the remaining elements of the regional system (TSs, VRFs, and IPCs). Finally, the remaining bulky waste landfills are examined separately. These landfills are remnants of the older municipal system for waste disposal. They are used as a proxy measure of the older system to determine whether newer facilities are more or less likely to be located in poor and/or racial minority areas.

DATA COLLECTION AND METHODS

As indicated above, previous quantitative studies of the locations of waste disposal sites have employed a variety of methodological approaches. Almost all of them employ an ecological correlation analysis, in which the units of analysis are geographic areas that are compared on the basis of social-demographic characteristics of individuals, families, or households within the unit. Four aspects of these studies are particularly important: the size of the geographic unit of analysis; the number and variety of variables measured for each geographic unit; the type of dependent variable examined and the statistical analyses used; and how "comparison" geographic units were defined and used in the statistical analyses. We discuss below how our study compares on each of these issues.

Some studies have employed fairly large geographic units, such as zip code areas, towns, or even counties, as their units of analysis. These larger units have some disadvantages. First, they tend to be less homogeneous than smaller units, such as census tracts or block groups, and therefore do not provide good social-demographic profiles of local populations. In addition, describing the characteristics of populations across larger geographical areas increases the risk of committing the ecological fallacy. For example, suppose two studies reported that the geographic units with a high percentage of minorities are more likely to have landfills, but one study used counties and the other used census block groups. With the county data, there is less certainty that the minorities are actually located nearer the landfills. These uncertainties decline as smaller units of analysis are used. However, data availability problems start to appear with smaller units. For example, to protect the confidentiality of households reporting to the U.S. Census, some information is not reported at the census-block level. In addition, some variables might be missing at these smaller units; for example, the value of owner-occupied housing may be missing if all of the residents in a block area are renters.

We use block groups as our unit of analysis, dividing the state into 2,909 units. The land area for block groups is very skewed, with a mean of 1,130 acres and a median of 309 acres. The mean and median population sizes are 1,131 and 1,032, respectively. Our analysis includes selected variables created from the 1990 U.S. Census for each of these geographic units.

Number and Types of Variables Selected

In one sense, to assess environmental inequity we need only measures of race, ethnicity, poverty, and distance to facility for each block group. If we find that minorities or the poor are more likely to live closer to facilities, this would provide direct evidence that the burdens of environmental pollution are not shared equally among all social groups. The implications for social policies are quite different, however, if the facilities were placed originally in minority neighborhoods than if minorities arrived after the facilities were developed. Similarly, if the correlation between distance from facilities and race disappears when controlling for the effects of other variables, such as level of education, then this reduces support for racial injustice. The latter result might indicate that residents in areas with more educated members are better able to resist the placement of waste disposal facilities near them or are better able to relocate if a facility is sited in their neighborhoods. Thus, multivariate analyses can provide helpful, although not definitive, evidence for different explanations of observed bivariate patterns.[18]

We used or created a number of variables from the 1990 census that might provide likely explanations for the location of waste facilities.[19] For instance, besides measures of race, ethnicity, poverty, and income, we also included measures of education, industrial areas, housing and social density, and neighborhood stability. If we find—after controlling for these other variables—that race/ethnicity and/or poverty remains correlated with distance from a facility, we have a stronger basis for raising concerns about environmental injustice. We provide below a brief description of each variable.

%MINORITY: Non-Hispanic whites are the "majority" and all other racial or ethnic groups are classified as minorities. In Connecticut, "minority" includes mostly African American and Hispanic groups.

%POVERTY: The percent of persons living below the 1989 federally established poverty line.

INCOME/CAP: Per capita income (in thousands) for those individuals over fifteen years of age.

%BA+DEG: The percent of residents twenty-five years and older with a bachelor's degree or higher.

HOMEVALUE: The median value (in thousands) of owner-occupied homes.

HOUSINGAGE: The median age of the housing units in 1990.

%SAMEHOME: The percentage of residents living in the same home for at least five years.

%MANUFACT: The percentage of workers employed in the durable and nondurable manufacturing sector.

CHEMSPILLS: The number of chemical or oil spills per acre of land. This measure is based upon the CT Department of Environmental Protection's geographical information system (GIS) database of spills or leaks.

HU/ACRE: The number of housing units per acre.

PERSONS/HU: Average number of persons per occupied housing unit.

TOTALAREA: The total size of the block group (in acres). Urban block groups tend to be smaller and more densely populated.

%WATERAREA: The percentage of the unit area that is water.

Dependent Variables and Methods of Analysis

Previous studies have employed a variety of dependent variables and statistical techniques. Some have used a simple t-test to compare the features of the geographic units "hosting" facilities to those units adjacent to them (or to units in the same metropolitan area). The more sophisticated analyses used logistic regression to make multivariate comparisons between geographic units with (or near) facilities to those further away.

Our dependent variable is the distance in miles between the center, or "centroid," of the block group and the nearest disposal facility. There are several advantages to this strategy compared to the host/nonhost approach. First, since the effect of a facility does not stop at the boundary of the geographical unit of analysis, we have better control by using distance itself. Second, the different sizes of block groups can create problems in some settings. For example, rural block groups are much larger than urban block groups, so a contrast of host versus nonhost block groups can involve much larger distances in the rural setting. Similarly, a large host block group might have a facility located at its edge, where a small adjacent block group's centroid would be closer to the facility than the centroid of the larger host block group. These possibilities suggest it would be better to explicitly measure distances as directly as possible. Finally, by using distance as a quantitative dependent variable, we can use regression analysis with its easy interpretations of the effects of the predictor variables.

Comparison Groups

Research results may depend on how "comparison" units are defined. In a logistic analysis, the comparison group defines which geographic units without facilities are contrasted with those with facilities (or near facilities). For example, the comparison group might be all other units without facilities but in the same town, SMSA, county, state, or nation. One of the criticisms that has been made about environmental equity studies is that they assume that a facility could be located anywhere within a large geographical area—a state for instance. However, some areas might not be feasible locations because they are not located close to transportation routes, do not meet geological requirements, or are not located close to populations where waste is generated.[20]

Since distance from the facility is our dependent variable, we use it to define our comparison group. Our analysis uses block groups within a ten-mile radius of a facility for several reasons. First, it is reasonable to assume that locations within ten miles are more feasible alternative sites than any location within the state, county, or metropolitan area. Choosing an alternative location within ten miles of a facility is not likely to increase transportation costs by much, should still locate the facility close to consumer populations, and is less likely to provide radically different geological conditions. Second, a ten-mile radius should also provide enough variation in population characteristics to make our analysis meaningful. Finally, since the number of comparison units will increase as a square of the distance (assuming geographic units tend to be the same size), without some mileage limit the comparison group may become so large and diverse that the more refined contrasts within ten miles are "washed out" by the features of the distant and more numerous block groups. We thought it was more important to capture variation in population characteristics in areas more immediately around facilities, where reasonable alternative locations might be more plausible. Thus, we restricted the regression analysis to those block groups within ten miles of the current facilities. Table 4.1 shows the number of sites for each type of facility and the number of block groups located within ten miles of three groupings of facilities: the heart of the regional waste disposal system (RRFs and their ash landfills), the entire regional system, and the bulky waste landfills.

Table 4.1. Types of Solid Waste Disposal Facilities by Number of Sites and Number of Surrounding Block Groups

	Regional Facilities					*Local Facilities*
	RRF	*Ash LF*	*TS*	*VRF*	*PC*	*BWLF*
# Sites	7	6	88	13	6	73
# Block Groups within 10 Miles		—————	2000	————————————————	2,821		2,769

ANALYSIS

Figure 4.1 provides a broad overview of the bivariate relationships between the distances to each type of MSW facility and the percent minority. The vertical axis gives the mean distance to the nearest facility of each type for ten sets of block groups, which are arrayed along the horizontal axis based upon their percent minority. All block groups with 0–10 percent minority are given the value of the midpoint or 5, those with a score of 15 have 10–20 percent minority, and so forth.[21] This figure indicates that the block groups with more minorities are generally closer to the RRFs, IPCs, VRFs, and ash landfills. On average, the block groups with 90 percent or more minority members are about five miles closer to these facilities—or about half the distance—than the block groups with less than 10 percent minorities. The remaining component of the regional system is the set of transfer stations, which are more numerous but show almost no relationship with the percent minority. The older bulky waste landfills, which are taken as continuing remnants of the old town-based disposal system, show a pattern similar to the transfer stations. This consistency between these two types of facilities is not surprising, because most transfer stations are located at closed landfills or at a few of the current bulky waste fa-

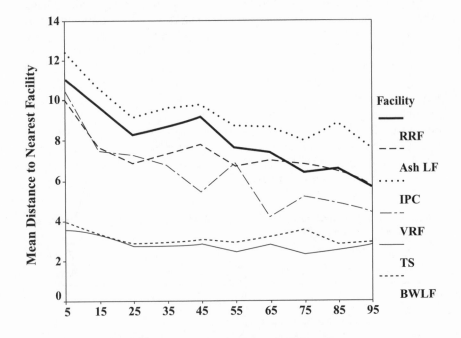

Percent of Block Group That Is Minority

Figure 4.1.   Mean Distance to Type of MSW Facility by Percent Minority

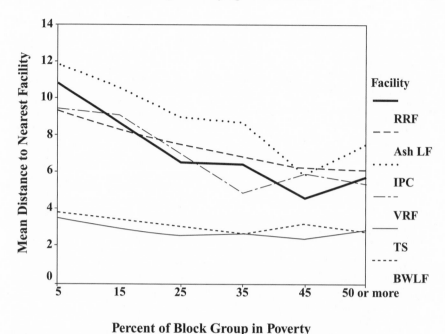

Percent of Block Group in Poverty

Figure 4.2.   Mean Distance to Type of MSW Facility by Percent in Poverty
Note: The 29 block groups in "50 or more" include 10 with >60% poverty.

cilities. Figure 4.2 provides similar information, based upon the percent of persons in poverty in the block groups. It shows changes that are very similar to those shown in figure 4.1, but with slightly smaller declines in average distance to facilities as the percent in poverty increases.

These bivariate plots indicate that block groups with the highest percentages of minorities or the highest levels of poverty are generally closer to most elements of the regional municipal solid waste disposal system. This association should raise some concern about environmental equity issues, especially if these relationships remain after controlling for the effects of other variables.[22] To test whether these relationships are spurious, we used the previously described variables in a regression analysis predicting the distances to the nearest facility within various facility groupings: (a) the "burn and bury" heart of the regional system (the RRFs and their ash landfills), (b) all components that made up the regional system of handling municipal solid waste in the mid-1990s (see table 4.1 for a breakdown by type), and (c) the seventy-three bulky waste landfills remaining from the previous town-based disposal system.

Table 4.2 gives the descriptive statistics for the block groups within ten miles of each of these three groupings of facility types. The mean scores for most variables are remarkably similar across these different groupings, even though the number of block groups in the groupings vary and the focal facility types

74

Chapter 4

Table 4.2. Descriptive Statistics of Block Groups around Different Types of Solid Waste Disposal Facilities

Variables LF	Type of Facility		
	RRF & Ash LF	Regional	Bulky Waste
Distance in Miles	5.38	2.55	3.54
	2.58	1.49	2.20
%Minority	19.54	15.29	15.51
	26.39	23.51	23.67
%Poverty	7.73	6.65	6.67
	10.67	9.49	9.55
Income/Cap	20.37	20.77	20.84
	11.08	10.84	10.92
%BA+Deg	26.31	27.27	27.43
	18.27	17.49	17.55
%SameHome	58.00	58.26	58.22
	15.22	14.53	14.58
HomeValue	185.4	188.2	188.8
	91.55	87.15	87.62
HousingAge	34.38	33.05	33.05
	12.25	12.39	12.36
HU/Acre	.798	.643	.648
	.937	.849	.854
Persons/HU	2.62	2.63	2.63
	.428	.418	.418
%WaterArea	3.34	4.41	4.46
	10.79	12.98	13.09
TotalArea (acres)	693	1143	1133
	1413	2165	2156
%Manufact	19.99	20.41	20.32
	8.84	8.74	8.71
ChemSpills	.0015	.0013	.0013
	.0049	.0044	.0043
Sample Size (N)	2000	2821	2769

Note: The means and standard deviations (above) are given for each combination of variable and type of facility. Block groups beyond 10 miles are excluded from the statistics for each type of facility.

are different. This indicates that the ten-mile comparison groups are similarly diverse, which reduces the likelihood that our later results are influenced by differences in the comparison groups. The largest difference is in the means for the total area of the block groups. The much lower means for the RRFs and their ash landfills reflect their locations in more urban surroundings. However, the interesting differences between these groupings appear in the regression results given in table 4.3.

Table 4.3 presents both the unstandardized and standardized regression coefficients for all the variables described in the methods section.[23] The most

Table 4.3. Predicting a Block Group's Distance to Different Types of Solid Waste Disposal Facilities

Variables	Type of Facility		
LF	RRF & Ash LF	Regional	Bulky Waste
Constant	4.97***	2.40***	2.28***
%Minority	−.0076*	−.00969***	−.00632*
	−.078	−.153	−.068
%Poverty	.0146*	.00917	−.000033
	.061	.058	−.000
Income/Cap	−.00692	.0040	.00674
	−.030	.029	.033
%BA+Deg	.0353***	.0168***	.00638
	.251	.197	.051
%SameHome	−.00288	.00107	−.00627
	−.017	.010	−.042
HomeValue	−.00290*	−.00328***	.000734
	−.103	−.192	.029
HousingAge	−.00495	−.0112***	.00362
	−.024	−.093	.020
HU/Acre	.0608	−.167***	−.0832
	.022	−.095	−.032
Persons/HU	−.0928	.237**	.348**
	−.015	.066	.066
%WaterArea	.00118	.00240	−.0118***
	.005	.021	−.070
TotalArea (acres)	.000229***	.000108***	.000133***
	.125	.157	.130
%Manufact	.0289***	−.000409	.0104
	.099	−.002	.041
ChemSpills	−5.27	−16.69**	−11.8
	−.010	−.049	−.023
R-Square (Adj.)	.044	.152	.050
Sample Size (N)	2000	2821	2769

Note: Block groups beyond 10 miles are excluded from the analysis for each type of facility. The standard-ized regression coefficients are given below the unstandardized coefficients.
* Significant at the .05 level or better.
** Significant at the .01 level or better.
*** Significant at the .001 level or better.

important result is the significant impact of %MINORITY in all three equations. Furthermore, it is important to note that %POVERTY is significant in only one equation and is not in the predicted direction: block groups with more people in poverty tend to be further away from RRFs and their ash landfills. Thus, race/ethnicity continues to be a significant predictor of distance from a facility with a relationship that supports concern about environmental equity issues and raises the question: How important is this effect?

The percent minority ranges from 0 to 100 percent among the block groups. The unstandardized coefficients indicate that the average effect of moving from 0 to 100 percent minorities will be about 0.76, 0.97, and 0.63 miles closer to, respectively, the RRFs and their ash landfills, all the components of the regional system, and the bulky waste landfills. These shifts appear quite modest, but they might be large in a relative sense; for example, a one-mile shift is large in a relative sense if it is from 1.5 miles to .5 miles (or a 67 percent decrease in distance from the facility). The largest relative shift is for the regional facilities, where the .97-mile shift moved the 100 percent minority block group about 36 percent closer to the facility.[24] Finally, the standardized coefficient for %MINORITY (-0.153) in the regional equation indicates that its effects are stronger in the new regional system than in the older system indicated by the analysis of the bulky waste landfills, where %MINORITY's standardized coefficient is only -0.068.

These results suggest that regionalization has increased the proximity of minorities to the various components of the municipal solid waste disposal system. Some of our unreported analyses suggest that this occurred by bringing urban block groups with higher concentrations of minorities into closer contact with this system. The older system utilized landfills dispersed in rural settings, which had lower concentrations of minorities, whereas the addition of urban RRFs, IPCs, and even ash landfills has increased the proximity of urban minorities to the solid waste disposal system.[25]

The comparisons of effects of percent minority with other variables within each equation show some interesting contrasts. For the analysis concerning the regional facilities, the effects of education are much stronger than the effects of percent minority. Block groups with higher %BA+DEG scores are more likely to be further away from RRFs and ash landfills or the combined regional facilities. This may reflect that more educated residents have both the economic and political capital to resist or avoid these facilities.

Aside from %MINORITY, only one other variable has a significant effect in all three equations: the TOTALAREA of the block group. Its strong and positive effect indicates that larger block groups are more likely to be distant from facilities of any type. Area was included as an important statistical control: the distance of a block group was calculated from its center, so larger block groups must be further away on average than smaller ones. Since urban block groups are smaller than rural ones, it is possible that the positive effect of area might also indicate a tendency for urban location of the facilities. However, other variables may be better measures of this aspect, such as housing units per acre (HU/ACRE) and median age of housing units (HOUSINGAGE). For the regional facilities as a whole, these two variables do have significant negative effects, which indicates that housing units near regional facilities are older and more densely located—both indicators of urban settings. However, the strongest predictor for regional facilities as a whole is the negative effect of median value of owner-occupied housing (HOMEVALUE), which suggests that facilities are closer to more expensive homes, which often have more rural locations. Unreported separate analyses by specific facility type indicate that the negative effect of HOMEVALUE occurs only for the locations of transfer stations and some ash landfills, which are more rurally located.

Three other variables reach significance in various equations. First, we find that the percent of residents employed in the manufacturing sector (%MANUFACT) has a positive correlation with distance from an RRF or ash landfill. This opposes the pattern identified by Anderton and his colleagues, who found that tracts with many skilled and semiskilled operators were more likely to have commercial hazardous waste facilities, which they took as indicating a preference to locate these facilities in industrial areas. Our results may differ because we constructed our variable differently, using the percent of the employed who are working in economic sectors that manufacture durable or nondurable goods, which would include managers. In addition, we also used different statistical techniques and comparison groups. However, in the regional equation, we do find results consistent with the Anderton interpretation. Block groups with many chemical and oil spills (CHEMSPILLS) are more likely to be closer to facilities and these spills are probably a more direct indicator of industrial activity than of the types of workers living in the block group. Finally, the equation predicting the location of bulky waste landfills indicates that they are more likely to be located near block groups with a higher percentage of their area in water (%WATERAREA). This might indicate that these older landfills were often placed in "less valuable" lowlands and swamps.

CONCLUSION

It appears that the initial bivariate relationship between proximity to regional solid waste facilities and the percent poor is spurious, whereas the relationship with the percent minority remains in the presence of substantial statistical controls. The effect of percent minority is the highest in the equation for all regional facilities combined. Other variables, such as education, home value, and area, have equal or stronger effects, as reflected by their standardized coefficients, but the continued impact of percent minority should raise concerns about the impact of regionalization on environmental equity.

The results here provide only a snapshot of a dynamic process. Subsequent research should examine more closely the longitudinal development and consequences of the regionalization process. Adding census data from 1980 and 2000 will increase our ability to discern cause and effect. For example, it would be possible to see if the regional facilities built in the 1980s caused a decline in the surrounding property values between 1980 and 1990 and whether this was followed by an increase in the percent that are poor between 1990 and 2000. Another type of study of the causal process could use the fact that the majority of current, regional facilities are located at old, closed landfills. An analysis contrasting the traits of these reused landfill locations with those that were simply closed might help illuminate the causal processes producing Connecticut's present system.

There are also many ways that these regression analyses might be improved. One key improvement would take into account two important features of the facilities: pollution volume and pollution diffusion with distance. For example, the urban RRFs in Connecticut burn four to five times as much trash as some of the smaller, rural RRFs. Although the pollution volumes released from the smokestacks of the RRFs depend strongly on the technologies used, the pollution from the diesel trucks carrying the trash is a known carcinogen and the volumes released will be proportional to the volumes of trash burned. Furthermore, truck pollution will be released at ground level. This pollution will diffuse as a nonlinear function of distance, so that pollution levels at ? mile will be more than twice those found at ? mile. Since minorities are much more concentrated in urban areas, both of these modifications could have a dramatic impact on the results. Taking into account wind direction would also be an important addition to the diffusion model.

Finally, subsequent research might explore other modifications of the approach used here. For example, adding variables for distinct minority groups, for example, African Americans and Hispanics, might produce different results. Exploring interactions among the predictor variables is another possibility, because it seems quite possible that the effects of being poor *and* black are likely to be more than the sum of the effects of each variable considered separately. Another modification would be to examine whether changing the comparison group from block groups within ten miles to those within, say, five miles makes a significant difference in the results.

Although these suggestions presume a quantitative approach to the study of environmental justice, we firmly support qualitative approaches as well. Indeed, the best chance for advancing the field involves combining the insights from both approaches and encouraging a debate among researchers, policy makers, and public citizens.

NOTES

An earlier version of this paper appeared in *The New England Journal of Public Policy* (Spring/Summer 2001). Permission to reprint was granted by the journal's editor.
    1. John Cushman Jr., "Pollution Policy Is Unfair Burden, States Tell E.P.A." *New York Times*, May 10, 1998, A16.

2. United Church of Christ, Commission on Racial Justice, "Toxic Wastes and Race in the United States: A National Report on the Racial and Socioeconomic Characteristics of Communities with Hazardous Waste Sites" (New York: United Church of Christ, 1987).

3. Douglas L. Anderton et al., "Hazardous Waste Facilities: 'Environmental Equity' Issues in Metropolitan Areas," *Evaluation Review* 18, no. 2 (April 1994): 123–40.

4. Douglas L. Anderton et al., "Environmental Equity: The Demographics of Dumping," *Demography* 31, no. 2 (May 1994): 229–48.

5. Andy B. Anderson, Douglas L. Anderton, and John Michael Oakes, "Environmental Equity: Evaluating TSDF Siting over the Past Two Decades," *Waste Age* 25, no. 7 (July 1994): 83–100.

6. Anderton et al., "Hazardous Waste Facilities."

7. Evan Ringquist, "Equity and the Distribution of Environmental Risk: The Case of the TRI Facilities," *Social Science Quarterly* 78, no. 4 (December 1997): 811–29.

8. John Hird and Michael Reese, "The Distribution of Environmental Quality: An Empirical Analysis," *Social Science Quarterly* 79, no. 4 (December 1998): 693–716.

9. Francis O. Adeola, "Environmental Hazards, Health, and Racial Inequity in Hazardous Waste Distribution," *Environment and Behavior* 26, no. 1 (January 1994): 99–126; Douglas L. Anderton, John Michael Oakes, and Karla L. Egan, "Environmental Equity in Superfund: Demographics of the Discovery and Prioritization of Abandoned Toxic Sites," *Evaluation Review* 21, no. 1 (February 1997): 3–26; Douglas L. Anderton, John Michael Oakes, and Michael R. Fraser, "Race, Class and the Distribution of Radioactive Wastes in Massachusetts," *New England Journal of Public Policy* 15, no. 1 (Fall/Winter 2000): 79–96; William M. Bowen, Mark J. Salling, Kingsley E. Haynes, and Ellen J. Cyran, "Toward Environmental Justice: Spatial Equity in Ohio and Cleveland," *Annals of the Association of American Geographers* 85, no. 4 (December 1995): 641–63; Michael Gelobter, "Toward a Model of 'Environmental Discrimination'," in *Race and the Incidence of Environmental Hazards*, ed. Bunyan Bryant and Paul Mohai, 64–81 (Boulder, Colo.: Westview, 1992); Eric J. Krieg, "The Two Faces of Toxic Waste: Trends in the Spread of Environmental Hazards," *Sociological Forum* 13, no. 1 (March 1998): 3–33; Paul Mohai and Bunyan Bryant, "Environmental Racism: Reviewing the Evidence," in *Race and the Incidence of Environmental Hazards*, ed. Bryant and Mohai, 163–76; Phillip H. Pollock III and M. Elliot Vittas, "Who Bears the Burdens of Environmental Pollution? Race, Ethnicity and Environmental Equity in Florida," *Social Science Quarterly* 76, no. 2 (June 1995): 294–310; Paul Stretesky and Michael J. Hogan, "Environmental Justice: An Analysis of Superfund Sites in Florida," *Social Problems* 45, no. 2 (May 1998): 268–87; Harvey L. White, "Hazardous Waste Incineration and Minority Communities," in *Race and the Incidence of Environmental Hazards*, ed. Bryant and Mohai, 126–39; Tracy Yandle and Dudley Burton, "Reexamining Environmental Justice: A Statistical Analysis of Historical Waste Landfill Siting Patterns in Metropolitan Texas," *Social Science Quarterly* 77, no. 3 (September 1996): 477–92; Rae Zimmerman, "Social Equity and Environmental Risk," *Risk Analysis* 13, no. 6 (December 1993): 649–66.

10. Michael Greenberg, "Proving Environmental Inequity in Siting Locally Unwanted Land Use," *Risk: Issues in Health and Safety* 14, no. 3 (Summer 1993): 241.

11. Robert D. Bullard, "Solid Waste Sites and the Black Houston Community," *Sociological Inquiry* 53, nos. 2–3 (Spring/Summer 1983): 273–88.

12. Vicki Been, "Locally Undesirable Land Uses in Minority Neighborhoods: Disproportionate Siting or Market Dynamics?" *Yale Law Journal* 103, no. 6 (April 1994): 1383–1411.

13. Been, "Locally Undesirable Land Uses," 1390.

14. Greenberg, "Environmental Inequity."

15. General Accounting Office, "Hazardous and Nonhazardous Waste: Demographics of People Living Near Waste Facilities" (Washington, D.C.: General Accounting Office, 1995).

16. William Markham and Eric Ruffa, "Class, Race, and the Disposal of Urban Waste: Locations of Landfills, Incinerators, and Sewage Treatment Plants," *Sociological Spectrum* 17, no. 2 (April–June 1997): 235–48.

17. Markham and Ruffa, "Class, Race," 238.

18. Longitudinal data sets would also be helpful for understanding the social-historical dynamics behind the observed patterns.

19. The data source was GeoLytics, Inc.'s "Census on CD" product, which is based upon the 1990 Summary Tape File 3 (STF3) census data set. The STF3 data are based upon the "long" census form that is only given to a sample of the population, however the STF3 data have been weighted to represent the total population.

20. Richard D. Clark, Steven P. Lab, and Lara Stoddard, "Environmental Equity: A Critique of the Literature," *Social Pathology* 1, no. 3 (Fall 1995): 253–69.

21. In these bivariate figures, we use all block groups in the state to view the larger pattern and, in a sense, to establish a baseline before conducting the multivariate analysis, where the analyses are restricted to block groups within ten miles of a facility.

22. Even if the bivariate relationships are found to be spurious, they would still be a source of concern. Distributions of racial minority and poor populations may turn out to be poor predictors of facility locations, but if facilities are disproportionately located in these communities, they will remain public concerns. Socioeconomic processes may produce these bivariate correlations or they could result from unintentional consequences of siting criteria rather than as intentional consequences.

23. We included all variables for two reasons. First, some readers may want to know the effect of a particular variable, even if it is not statistically significant. Second, one might argue that we are studying the entire population of block groups, so significance tests are irrelevant. We did do a backward elimination of less significant predictors and obtained essentially the same results. Furthermore, weighting block groups by the number of residents yielded very similar patterns of influence, but all variables became statistically significant with the increase in "sample" size to over three million.

24. In order to calculate relative shifts, an initial distance was calculated for the "average" block group with no minorities. This was done by using the means of all the predictor variables in the regression equation, except %MINORITY was set to zero. Then the effect of a change in percent minority from 0 to 100 percent was calculated and used to determine relative change.

25. The reported analyses represent an aggregated result. Each separate facility site could be analyzed in the same way as reported here, but we aggregated by type

of facility. We did discover that the effect of %MINORITY varied drastically if we separated urban sites from rural sites. For rural sites, there was a strong positive effect for %MINORITY, indicating that minorities were concentrated in the more distant urban areas. Just the opposite occurred for facilities located in urban settings. We do not report these separate results because we want to assess the *overall effects* of the total regional system and its various components.

5

Risky Business? Relying on Empirical Studies to Assess Environmental Justice

Pamela R. Davidson

Concerns for environmental justice have generated an impressive body of empirical research over the past fifteen years. The conclusions of this research are far from consistent. On the contrary, results vary widely and are even contradictory. There appears to be a basic divide in the field of environmental justice between researchers who present evidence in support of environmental inequities and those who conclude that there is little evidence of widespread environmental injustices. However, this divide may only be a methodological artifact, since the methods researchers use to make their claims in one direction or the other vary widely.

Researchers have been quick to point to methodological differences as the cause of inconsistent and contradictory findings inherent in the field of environmental justice research. This discourse has been useful to the field. As environmental justice research has matured, practical and methodological limitations to the assessment of inequities have become evident. Over the years, a number of methodological issues have been discussed, while others have received little attention. In this chapter, I draw attention to some of the neglected or implicit methodological challenges. The most basic point that this chapter makes is that a fundamental challenge facing environmental justice researchers is the establishment of clarity regarding the research hypothesis. Directly related challenges include an understanding of the usefulness of various units of analysis and research designs in investigating a research question that is unambiguously defined. Finally, as environmental justice research shifts focus from siting disparities to impact disparities, it faces new methodological challenges, some of which are laid out below.

AMBIGUOUS HYPOTHESES:
DECIDING ON THE RESEARCH QUESTION

One of the challenges facing environmental justice researchers is resolving the mismatch between the research hypothesis and research tools. Environmental justice research has always implicitly been about the distribution of "environmental risk" without necessarily explicitly quantifying risk in epidemiological terms. In recent times, environmental justice researchers, on the basis that proximity is a poor proxy for risk, have become critical of studies focused on the siting process.[1] Even when the critically scrutinized studies do not claim that proximity is a proxy for risk, the fact they are criticized on this basis does provide an indication of the lack of clarity regarding the research question in environmental justice research. Simply stated, in environmental justice research, is the research question about siting biases (e.g., the decision to locate a hazardous waste treatment facility in a poor or minority community) or about risk-based disparities (e.g., minority communities bearing a disproportionate share of health risks potentially related to proximity to hazardous waste sites)?

It is probably best to begin by looking at what the two hypotheses are and why the distinction is relevant. In environmental justice research, the hypothesis of discriminatory siting is that noxious environmental sites are disproportionately located in poor and minority neighborhoods because they are either least likely to prevent a siting (political empowerment thesis)[2] or because factors such as cheaper housing or racial discrimination propel them into such areas (market forces thesis).[3] The hypothesis of disparate risk is that low-income and minority populations bear a disproportionate share of environmental risks for two reasons.[4] First, they are more likely to be exposed to multiple types and sources of environmental toxins in their neighborhoods, jobs, and cultural customs and, second, due to differential health risk factors, they are more susceptible to environmental exposure. Naturally, many environmental justice researchers would be quick to point out that environmental justice research is not just about a single issue. In fact, Robert Bullard argues explicitly, "Environmental justice is not just about facility siting. It also involves issues and concerns around pesticide exposure, lead poisoning, transboundary toxic waste dumping, shipping risky technologies abroad, unequal protection, differential exposures, and unequal enforcement of environmental, public health, civil rights, and housing laws."[5]

Since environmental justice researchers may have multiple concerns and view inequities on a number of levels, this complexity may translate into ambiguity and methodological weaknesses. The confusion and ambiguity visible in the past essentially stems from two sources. First, it results from using an indicator that could serve as a proxy either for discriminatory siting or disparate risk, but in each case is a weak indicator. This indicator is spatial coincidence or the presence of hazardous waste sites in the same areas in which there are large numbers of minorities or low-income residents. Second, it results from using unclear language regarding the research hypothesis in which the researcher toggles back and forth between emphasizing bias and emphasizing risk. The problem with this is that areas of risk and impact may be very different from areas we might designate as "targeted neighborhoods."

While it is important to maintain clarity regarding the research question, both research thrusts are relevant to the investigation of environmental inequities. Nonetheless, they are distinct hypotheses and require different research strategies. The distinction between discriminatory siting and disparate risk is particularly salient for environmental justice research because of differences in findings that seem to be related to the size of geographic areas serving as the unit of analysis. As research is beginning to reveal, the demographics of host areas may not be representative of the demographics of the areas of impact.[6] While smaller areas such as U.S. Census tracts may be more appropriate for testing the siting hypothesis, it is not yet clear which geographic unit is appropriate when testing for the hypothesis of disparate risk. The evaluation of risk is complex, since it must factor in exposure mediums, cumulative risk, toxic synergies, geological landscape, population sensitivity, and other factors, all of which make it inherently more difficult to study. Since environmental justice research has found that inequities along racial and ethnic lines become more apparent the larger the unit of analysis, it is plausible that risk-based research may produce very different findings than research based on the siting hypothesis. In particular, if larger geographic areas are more relevant to the risk hypothesis, then stronger evidence for environmental inequities may be visible for the disparate risk hypothesis than has been visible in environmental justice research thus far.

Size Matters: The Relevance of the Unit of Analysis to the Research Question

An unresolved issue in environmental justice research is the appropriate unit of analysis. Early studies finding evidence of a disproportionate representation of disadvantaged populations in host areas used zip codes or counties, while later studies with the opposite findings relied on tracts or block areas.[7] The use of different units of analysis provides one explanation for contradictory research findings regarding the relationship between the presence of environmental hazards and the demographic characteristics of areas, since correlation tends to increase with aggregation.[8] Anderton et al. attempted to reconcile these findings by demonstrating that, by aggregating tracts hosting hazardous waste facilities with other tracts within a 2.5 mile radius, similar findings suggestive of environmental inequities along racial lines emerge.[9] These issues of spatial data resolution have been reviewed in greater detail elsewhere and suggest that, while hazardous waste sites may not be in the backyards of minorities, minority neighborhoods may indeed be at least in the same larger geographic areas (e.g., counties) as hazardous waste sites.[10] An additional issue that is often overlooked is the importance of using the unit of analysis that best suits the stated research goal.

If the stated purpose of the research is to investigate discriminatory siting and distributional disparities in the location of industrial hazards, then positive findings based on aggregated areas such as counties or zip codes provide only weak evidence, since these geographic units can hardly be considered "targeted neighborhoods." With large aerial units, a positive relationship between race and facility location could describe a situation in which facilities are located in a large county, such as Los Angeles County, that contains a large minority population, but are still miles away from minority neighborhoods. For example, in his study of noxious industries in the Boston area, Eric Krieg found that industrial firms tend to "leapfrog" neighborhoods, creating a checkerboard of economic opportunity and decline as these firms followed whites out of central cities and into suburbs.[11] The result is that in newer areas in which industrialization coincided with suburbanization, industries and environmentally sensitive sites are located *in* white neighborhoods, but only *near* black neighborhoods. In contrast, in older industrial areas, race was found to be more strongly correlated with toxic waste site location. In the Boston area, as in many other large cities across the United States, residential and workplace discrimination has kept blacks out of the areas of commercial activity characterized by employment in the craft sector in a manner that would not be visible at a more highly aggregated level. These larger social forces have created a "lumpy" geography, in particular in urban settings, with dense clusters of ethnic and low-income neighborhoods appearing in "hypersegregated," checkerboard space.[12]

In these settings, the demographics of counties or zip codes in which environmentally sensitive sites appear may differ from the demographics of host neighborhoods. In fact, the aggregate statistics used to describe host counties or zip codes may not be descriptive of a single neighborhood in that large geographic unit. By using large areas such as counties or zip codes, researchers risk promoting an ecological fallacy, that is ascribing aggregate characteristics to all individuals who form the aggregate. The potential for doing so increases as the geographic scale increases.[13]

Not all researchers support the use of tracts as the unit of analysis in the study of environmental justice. In his criticism of the use of tracts in environmental justice research, Paul Mohai argues that the tract level may be too small a unit, since the "adverse impacts from a local environmental hazard or unwanted land use may well extend beyond the unit's boundaries."[14] Similarly, more recent works in environmental justice call for the use of statistical methods that take spatial autocorrelation into account, since the impact of environmentally sensitive sites may not be confined to the area actually hosting the site. This, however, is a separate research question and provides an example of how the research question is often confused in environmental justice research. While it is deserving of equal attention and additional investigation, positive findings using larger geographic areas or techniques that are better equipped to measure the impact of sites would say very little about siting biases and disparities. Research that focuses on the impact of environmentally sensitive sites is concerned with issues of risk, exposure, and other negative externalities, such as declining property values, increased noise or stench, and faltering economic development. Here, the research question changes from asking whether minority and low-income neighborhoods are disproportionately targeted for facility siting (disparate siting) to asking whether minority and low-income populations are disproportionately impacted by the location of environmentally sensitive sites (disparate impact).

Changing the research question to a more risk-oriented view introduces other methodological issues that will be addressed below. The main point for now is that environmental justice research is not always clear in its purpose. Studies that aim to ascertain which neighborhoods are the "dumping grounds" for hazardous waste do not answer this question by discussing the impact of living proximate to the dumping grounds.

Natural Communities

Adding to the debate on the appropriate unit of analysis are researchers who argue that tracts can be too large.[15] Census tracts cover a smaller area and are more homogeneous in terms of population size than zip codes and counties, making tracts more appropriate for neighborhood-level analysis. For this reason, census tracts are used frequently in environmental justice research as a proxy for neighborhoods and even in court cases investigating issues of environmental justice.[16]

Despite the increased use of tracts in environmental justice research, there remain questions regarding the appropriateness of tracts as proxies for the "invisible" communities that host environmentally sensitive sites. For example, Bullard's research on Houston demonstrated that one 1970 census tract actually contained several racially distinct neighborhoods.[17] While the census tract used to make this demonstration appeared to be predominantly white, the neighborhood within that tract (Almeda Plaza) that was adjacent to two landfill sites was predominantly black. In this case, tract-level analysis failed to detect a relationship between race and landfill location. The recently released "Final Guidance for Incorporating Environmental Justice Concerns in EPA's NEPA Compliance Analyses" issues a similar caution pertaining to the use of census tracts, since they may miss certain "pockets" of minority populations.[18] Despite these potential shortcomings, more varied data are available for tracts than for blocks (a census enumeration unit smaller than tracts), making the former more interesting to environmental justice researchers. Moreover, racial segregation may reduce the significance of within-tract racial heterogeneity.[19]

Because census tracts may be either too large or too small, an alternative approach to defining neighborhoods consists of interviewing either community leaders or neighborhood residents themselves.[20] Interviews with community leaders might reveal the geographic units that leaders feel they are representing whose residents have a self-identity and a common stake. Interviews with residents should provide information on perceptions of community boundaries. Taken together, it is argued, researchers may come up with areal units that more closely resemble "real" neighborhoods. As demonstrated by Allen Goodman, however, it is precisely with defining boundaries that individuals have the most problems.[21] Because people tend to have "fuzzy" border problems, it is unlikely that areas contrived through qualitative interviews with local populations would be mutually exclusive.

Locally defined neighborhoods present other problems as well. It is likely that the selection mechanism used to generate these areal units will be biased through selection on the dependent variable if the perceived common stake is shaped in terms of proximity to environmentally sensitive sites (the dependent variable). This also presents problems in terms of selecting the appropriate comparison group areas. And, finally, it is doubtful that methods used to locate these neighborhoods could be replicated on a large scale, a fact reducing the reliability and generalizability of findings based on such units. It may be true, then, as Susan Cutter, Danika Holm, and Lloyd Clark point out, that "the most appropriate scale lies beyond our ability to manipulate statistical information."[22]

Quality of Research Design

Environmental justice researchers who test for disparities in the siting of environmental risks along racial, ethnic, and class lines usually rely on correlational analyses to assess the potential spatial coincidence of the pertinent demographics (e.g., percent African American) and environmental risks (e.g., toxic landfill). Criticism of this approach has not been absent and centers on the use of cross-sectional designs,[23] meaningful geographic units,[24] and appropriate comparison groups.[25] Some of these methodological issues have been addressed by others. In the following, I consider some of these issues in the context of the special challenges facing environmental justice research.

Longitudinal Research Design

A research agenda focused on investigating distributional disparities as opposed to disparate impacts is best served by a longitudinal research design.

Only an analysis over time can reveal whether market dynamics or initial discriminatory siting explain observed disparities.[26] Nevertheless, it should be noted that a longitudinal design alone is not sufficient. It is best accompanied by the use of quasi-experimental equivalent control groups that place any observed shifts into the larger historical context of change.[27] For example, John Oakes et al. found that increasing percentages of blacks in tracts, once they became hosts to treatment, storage, and disposal facilities (TSDFs), did not constitute a "victory" for the market dynamics argument because this was a trend visible in most industrial settings.[28] Although the increase in minority representation in host tracts was substantive, it took place independent of facility siting and was more in line with the growing diversity of the types of areas in which facilities are typically found. Thus, changes observed over time in host areas must be compared to changes occurring in other comparable areas, called "quasi-experimental equivalent control areas."

A number of cases studies examining the histories of cities provide more in-depth examinations of the dynamic relationship between community demographics and noxious industrial sites. If a disparate distribution of environmentally sensitive sites is visible, case studies prove more useful than national studies in providing reasons for it. For example, if longitudinal analysis reveals that a particular site contained an overrepresentation of minority populations prior to siting, the actual reason for that siting is still unknown. An investigation of the standards used to judge environmental hazards at the time of siting may reveal that shifting perceptions of risk may be of relevance. In other words, "siting" never really took place, since the land use was initially not considered "hazardous." In instances in which true siting decisions take place, they are frequently influenced by local zoning regulations. The involvement of the local community, state and federal permitting agencies, and economic development agencies in zoning decisions all help to shape the industrial landscape and may have more to do with bringing noxious industries into disadvantaged neighborhoods (e.g., the only parts of the city in which such sites are permissible) than discriminatory intentions of individual plant owners.

Longitudinal Endogeneity

While the effect of local-level actors should not be underestimated, it also should not be overestimated, since their actions are constrained by the social and economic forces of history. This fact has important implications for an understanding of the nature and causes of distributional disparities pertaining to environmentally sensitive sites. Growth and development in cities is largely guided by path dependencies in which the decisions made by developers and planners in earlier times constrain the choices available today. Once a community becomes an industrial community, path dependency or longitudinal endogeneity makes it difficult for that community to be anything but industrial. Various factors such as proximity to railroad lines or an available labor force may explain how the community initially attracted industry. Once the initial step is taken, others quickly follow to solidify a certain pattern that, once in place, becomes difficult to dispense with. Thus, once a factory is located in a particular community, other industries dependent on that industry may follow that in turn cause more industrial workers to locate in proximity. Zoning represents a key institution that helps to congeal certain conditions into place. Zoning regulations may solidify existing inequities, since permit granting is usually made on the basis of the "appropriateness condition."[29] If the land use (e.g., heavy industrial use) is consistent with past and existing uses, then the construction of a new facility will be more readily permitted than it would be in communities without this history. In this sense, history must be viewed as a "trap to prevent communities long overburdened by uses no other area would tolerate from escaping that particular part of their past, and from realizing their vision for a better future."[30]

The racial composition of residential neighborhoods is also shaped by path dependency such that racial discrimination not only results in racial segregation, but also, together with institutional practices such as the lending practices of banks, enforcement of zoning laws, and practices of insurance companies, works to maintain segregation over time. The path-dependent nature of facility location is critical to the study of inequities in siting. Environmental justice researchers whose research goals include the evaluation of siting biases would do well to consider the extent to which observed distributional disparities are attributable to longitudinal endogeneity. While case study research provides one way to accomplish this goal, the generalizability of nonrandom local area case studies is questionable. At the other extreme, sophisticated longitudinal research designs using national-level data represent the exception[31] and not the rule in environmental justice research. Where longitudinal data are available, a research design that makes use of first differences would be an effective way to address the longitudinal equity issue of path dependency. A first step in this direction has already been taken,[32] but only for a limited group of environmental industries (commercial treatment, storage and disposal facilities).

FROM SITING TO RISK: THE SHIFT IN EMPHASIS IN ENVIRONMENTAL JUSTICE RESEARCH

The field of environmental justice research is undergoing a transition from an emphasis on the distribution of environmental burdens to an emphasis on risk. The bulk of earlier environmental justice research was concerned with the nature of possible siting inequities, with some studies finding a notable correlation between proximity to toxic chemical polluters and marginalized social groups.[33] Research drawing on this perspective compares the demographics of areas hosting hazardous waste sites to nonhost areas to determine whether there is a disparate distribution along racial, ethnic, or class lines. The sites that have been studied include commercial treatment, storage, and disposal facilities (TSDFs); inactive or uncontrolled toxic waste sites regulated under the Comprehensive Environmental Response, Compensation, and Liability Information System Act and its 1986 Superfund Amendments and Reauthorization Act (CERCLA/SARA) and listed in the CERCLA Information System (CERCLIS) or the National Priority List (NPL or Superfund); facilities whose toxic releases are listed in the Toxic Release Inventory (TRI); and hazardous waste handlers regulated under the Resource Conservation and Recovery Act (RCRA) and listed in the RCRA Information System (RCRIS).[34] Despite the early consensus, close to two decades of environmental justice research has failed to produce conclusive evidence in support of the siting hypothesis, resulting in a divisive field of study. Part and parcel of the divisiveness is the debate over whether it is race and ethnicity,[35] class,[36] or economics[37] that drive siting disparities.

Although the debates remain to be definitively resolved, most environmental justice researchers can agree to at least three conclusions pertaining to the siting hypothesis: (1) environmentally sensitive sites are neither uniformly nor randomly distributed across society; (2) there are large-scale differences between host and nonhost communities in terms of socioeconomic characteristics; and (3) racial and ethnic disparities are apparent in some local area studies and become clearer the larger the areal unit under investigation.

Siting or Risk?

For grassroots activists concerned with potential threats to the health and safety of their communities associated with noxious environmental sites, issues of siting and risk go hand in hand. The very cohesion of the environmental justice movement is based on the common perspective of individual communities being targeted by polluting industries because of their presumed greater willingness to accept potential health risks in exchange for jobs and an improved infrastructure. What is frequently ambiguous, however, is a clear understanding of what the siting issue actually is and which activities are deemed environmental risks. Is the siting issue one of big industrialists targeting minority and low-income communities as locations for their noxious industries or one of white middle-class communities being better informed and more resistant to undesirable siting decisions? In talking about risk, what sites or activities are of greatest concern? Activists are frequently more concerned with sites that the public considers more serious, even if their actual number is so small that both exposure and potential risk may be limited.

As the environmental justice movement continues to infiltrate the public policy arena, references to siting and risk remain somewhat ambiguous. The public policy focus clearly emphasizes ensuring adequate protection for all population segments from siting inequities. Enjoining this focus is Executive Order 12898, signed by President Bill Clinton on February 11, 1994, with its underlying assumption that the siting process has a disproportionate impact on minority and low-income populations that needs to be addressed. While the order vaguely addresses issues of risk when it considers the "effects" on minority and low-income populations of living near certain types of facilities, it more strongly emphasizes siting in its mandating of the assembly of demographic profiles around facilities. To this end, federal agencies are instructed to "collect, maintain, and analyze information on the race, national origin, income level and other readily accessible and appropriate information for areas surrounding federal facilities that are . . . expected to have a substantial environmental human health, or economic effect on surrounding populations." While the desired demographic measures are clearly laid out, the values of risk assessment are not clearly stated. Instead, they are ambiguously described in terms of health *effects* and expected health or economic effects.

The requirement since 1998 that environmental justice be a formal criterion for the preparation of environmental impact statements (EIS) and environmental assessments under the National Environmental Policy Act (NEPA) means that the risk perspective is more firmly established in environmental justice policy today.[38] With this, the struggle for environmental justice has expanded its focus as a civil rights issue to include being a legitimate public health issue.[39]

A more scientifically grounded understanding of risk may have become more important to grassroots movements as well. Using Woburn, Massachusetts, as an example, Phil Brown argues that a growing number of grassroots movements utilize elements of "popular epidemiology" as an important activist strategy.[40] Communities at risk are increasingly more likely to make use of epidemiological science to produce their own scientifically grounded risk assessments, which they hope will strengthen their cases before public officials.[41] Related to the shift in public policy at the grassroots level, more recent works in environmental justice call for the use of statistical methods that expand the geographic unit of study to better account for the impact of the location of environmentally sensitive sites. It is argued that conventional correlational methods are unable to take into account the fact that the impact of such sites may not be confined to host areas. The growing relevance of such techniques clearly demonstrates how the research question in environmental justice research is changing from asking whether minority and low-income neighborhoods are disproportionately targeted for facility siting to asking whether minority and low-income populations are disproportionately impacted by the location of environmentally sensitive sites. This shift to a more risk-oriented view introduces new methodological issues, some of which will be addressed below.

Methodological Approaches to Measuring Disparate Risk

A new emphasis in environmental justice research appears to be emerging that is less concerned with distributional inequities than with their impact in terms of risk of exposure. Risk assessments estimate the probability that an individual will experience a health effect given that he or she was exposed to a particular chemical substance. While this approach may seem particularly appropriate for environmental justice research, the use of risk-assessment techniques to evaluate the impact of noxious sites on minority populations has been criticized in the past. Specifically, Bullard has expressed concern that traditional risk-assessment tools such as those used by the EPA are not sensitive enough to detect the greater risk of exposure faced by minorities and other disadvantaged populations.[42] The use of a single-substance methodology makes it impossible to assess the cumulative, additive, and synergistic effects of chemicals. This is a valid point that must be weighed by researchers interested in pursuing a risk-based focus. Before going into any further detail about the direction that researchers should take, let's first look at how environmental justice researchers are currently incorporating risk into their analyses.

Spatial Coincidence

The first strategy makes use of the crudest indicator of risk. This strategy consists of simply recasting geographic coincidence in enumeration units, such as tracts, zip codes, or counties, as a measure of potential exposure. This transforms proximity to noxious sites from a proxy for siting discrimination into a crude proxy for risk. Simply stated, a host tract or county is thought to describe an exposure site, while nonhost areas do not. In studies such as these, there is no attempt to actually quantify risk, and exposure to risk is often used interchangeably with exposure to negative effects, such as stench, noise, or lower land values.[43] Using spatial coincidence to hazardous waste sites as a proxy for risk is problematic for many reasons. First, it disregards differences between sites. As Evan Ringquist explains,

> If both the facility itself and its pollution control equipment are operating properly, the risk posed by the facility may be minimal. In this instance, simply living near a[n] RCRA or TRI facility does not appreciably elevate any resident's exposure to environmental risk. Clearly, however, larger or poorly run facilities pose more risk than do smaller or well-run facilities.[44]

Similarly, there are important differences in terms of risk presented by newer and modern facilities and older facilities that have not been required to upgrade their operations due to special "grandfathering" arrangements. Second, this is a dichotomous exposure measure. It unrealistically assumes a constant level of exposure within a tract that suddenly stops outside the boundaries of the tract. And third, the sizes of tracts and counties and most other enumeration units are always different, making it difficult to characterize the area of risk.

Spatial Buffers

The problem of differently sized enumeration units is addressed in the methodological strategy of using spatial buffers. Instead of measuring spatial coincidence in enumeration units such as tracts, zip codes, or counties, some studies capture geographic proximity by using proximity measures such as circular buffers or zonal calculations. These measures usually pinpoint the locations of hazards more accurately and standardize sizes of the areas around them (e.g., hazardous waste sites are the centroids of areas defined as one-mile radii around sites). Nevertheless, as in research based on the siting hypothesis, results are sensitive to the shape and size of the buffer. Just as with research based on standard enumeration units, these strategies are still problematic for an analysis of risk, because they constitute a dichotomous exposure measurement. As Eric Sheppard et al. explain, they "do not rigorously account for how potential exposure depends on the direction and nature of distance decay of the diffusion of toxic chemicals released in the atmosphere."[45] Plume dispersion models represent an improvement by taking toxicological characteristics of chemicals and atmospheric conditions into account. But they too make simplifying assumptions and it is too tedious to apply them on a large scale. Early attempts on a small scale suggest, however, that racial disparities in the distribution of risk become apparent in plume dispersion models, even where other models fail to detect such disparities.

Spatial Autocorrelation Techniques

A third method showing increased popularity is the use of spatial autocorrelation techniques. These methods move beyond simple correlational techniques and take into account the characteristics of contiguous areas such as tracts. These techniques are particularly useful when using data with standard enumeration units such as tracts, since there is typically not enough information available to determine where in the tract the site is located. This is problematic for a risk assessment since a site located near the boundary of a tract could present a greater risk of exposure to neighboring tracts than to the actual host tract. Spatial autocorrelation techniques are more effective tools for dealing with these types of boundary problems. Nevertheless, they continue to provide a very crude approximation of exposure to risk, since they fail to measure risk directly.

As an aside, spatial autocorrelation techniques may in some cases also be appropriate for testing one variation of the siting hypothesis. It makes little sense from a siting perspective to posit that a hazardous waste facility will be sited in an area because there is an above average proportion of blacks nearby in neighboring areas.[46] It does make sense, however, to look to contiguous tracts for the representation of industrially employed persons. Location theory suggests that industries seek out areas that are not densely populated and that have an abundant supply of available industrial workers. It is consistent with this reasoning that neighborhoods abutting potential host neighborhoods would be important in that they provide an adequate supply of blue-collar workers. As Mohai points out, a large majority of blue-collar workers commute out of their place of residence to go to their place of employment. From a siting perspective, the supply of available (e.g., unemployed) blue-collar workers in the region is just as relevant as the supply in the host neighborhood, making the use of spatial autocorrelation techniques appropriate to study the effect of blue-collar worker availability on siting decisions.[47]

Direct Measures of Risk

In attempting to further refine the research tools to more appropriately incorporate risk into environmental justice research, some researchers create toxicity indexes[48] or estimate the average risk posed by acute hazards.[49] Such indexes, by taking a number of risk factors into account, provide more information about the nature of environmental risks and their estimated potential impact on exposed populations. By making use of information on the toxicity of the chemical, the size of the area affected by a potential accidental release, the probability of an accidental release, and other factors that determine risks, risk-based measures of environmental equity can be very different from proximity-based measures in revealing inequities. There are a number of limitations to this approach, including simplifying assumptions about release medians, paths of exposure, geomorphic features, and susceptibility. These factors are the same as those identified by Bullard as critical for environmental justice research. And all of the methodological strategies fail to deal with spatial-temporal correlation. This is a methodology that still needs to be developed.

Methodological Limitations in Measuring Risk

With the shift in focus from the study of inequities in siting to the study of inequities in exposure or risk, environmental justice researchers face new methodological challenges. As traditional risk assessment methodologies begin to be applied to the study of environmental justice, care must be taken to deal with the many well-known methodological issues, including accounting for the effect of competing risks, cumulative burden, and population mobility on risk or exposure estimates.

Even before environmental justice research began to incorporate a risk perspective into its analysis, there was concern over whether traditional risk and exposure models used by the EPA were sensitive enough to detect the greater risk of exposure to environmental hazards faced by minority and other disadvantaged populations. Bullard, for example, argues that current methodologies are incapable of assessing the cumulative, additive, and even synergistic impact of chemicals.[50] This is because the use of a single substance methodology discounts the effects of multiple exposures. This limitation is more serious for risk estimates in minority and low-income populations, since exposure to multiple substances is thought to be higher there.[51] The focus on single substances also fails to detect the risks of synergistic effects—the interactive effects that occur when one substance is altered by another. Two substances that taken alone may be harmless might interact to form a toxic substance. The risk associated with that chemical mixture would then be underestimated. As is well known in the fields of public health and epidemiology, it is often difficult to separate the effects of exposure from the effects of competing risk factors that influence health. For example, minorities and other disadvantaged populations may be at greater risk when they are exposed above some level and they may be more susceptible to the effects of exposure. Competing possible risk factors include access to health care, occupation, sexual behavior, lifestyle, diet, stress levels, and knowledge about environmental health issues. Disentangling the causal relationship is particularly challenging when the population under study is exposed to a number of risk factors. In addition, it is extremely difficult to demonstrate a causal relationship between exposure to environmental hazards and disease unless there is a strong association or the population is large.[52] One way to avoid problems associated with competing risks is to restrict the study to a very limited outcome, such as congenital defects resulting from residential proximity to one particular environmentally sensitive site.[53] Nevertheless, the use of nonexperimental data means it is nearly impossible to completely rule out competing risk factors.

A special challenge is presented by the movement of populations in and out of areas of risk and exposure, since mobility weakens our ability to make definitive statements about the causal relationship. The effects of population mobility on causal validity was seen, for example, in the late 1970s, in research that purported to show that above-ground nuclear testing in Nevada increased the incidence of childhood leukemia in Utah. Upon further investigation by the Kennedy Commission, however, it was established that most of the current residents were not residing in the area at the time of the testing.

Risky Data?

To prove or disprove the disproportionate effect of siting on minorities and low-income populations, a large bulk of environmental justice research utilizes data on empirically rare yet particularly dreaded sites, such as TSDFs and uncontrolled or abandoned toxic waste sites. With the focus on environmental risks and potential exposure to environmental risks, these data sets lose their appeal, since they do not represent the more common environmental risks to which communities are exposed. In their place, data sets such as RCRIS and TRI, with their broader coverage of environmental hazards, will become increasingly important. Nevertheless, environmental data currently available and in use by environmental justice researchers are insufficient to allow for an unambiguous investigation of the relationship between environmental health risks and socioeconomic variables. Each database has problems in that the chemical information provided prevents the accurate estimation of hazard magnitudes from all risk sources.[54] The toxic wastes presumably available at uncontrolled hazardous waste sites categorized as Superfund candidate sites (CERCLIS) have not yet been assessed. Those listed on the National Priority List (NPL or Superfund sites) have been assessed, but the actual amounts of toxins at a given site are not available. RCRIS lists hazardous waste in terms of generation and treatment process and not by chemical. A similar problem applies to TSDF databases.[55] TRI data do contain information on the magnitude of chemicals released, but the magnitude says very little about environmental hazardousness. In addition, TRI only collects information on releases and is ill equipped to track important risk-reducing changes that lower toxic chemical use.[56]

In its first official report in 1992, the EPA Environmental Equity Workgroup reviewed existing data on the distribution of environmental risks across population groups. They concluded that improvements must be made in data collection procedures, since environmental health data are not collected by race and income and do not contain information on the health risks posed by multiple industrial facilities, cumulative synergistic effects, or multiple and different pathways of exposure. Six years later, in its 1998 *Final Guidance,* the EPA listed two promising data sets, the National Human Exposure Assessment Survey (NHEXAS), developed by the EPA's Office of Research and Development, and the National Health and Nutrition Examination Survey III, developed by the EPA's Office of Policy, Planning, and Evaluation, that may prove particularly useful for health assessment studies within the environmental justice framework. The usefulness of NHEXAS data has already been demonstrated in a regional analysis of exposure and environmental justice in EPA Region V.[57]

CONCLUSION

In focusing on siting biases and distributional disparities, environmental justice researchers have failed to produce clear and consistent evidence. As such, there is no conclusive evidence of a widespread targeting of racial and ethnic neighborhoods for the siting of noxious environmental sites. Despite general disagreement in the field, most researchers agree that environmentally sensitive sites are not simply randomly distributed across the United States and that there are large-scale socioeconomic differences between host and nonhost areas. In addition, there is a general consensus that racial and ethnic disparities are visible in some local area case studies and that they become clearer in national studies the larger the geographic unit under investigation.

One major challenge facing researchers operating from a siting perspective is to resolve the ambiguity pertaining to the research question. The hypothesis of discriminatory siting is separate from the hypothesis of disparate risk, but the two are often used interchangeably, thereby affecting the validity of the research findings. An ambiguous research question may also translate into a weak research design in terms of choice of the unit of analysis. In general, tracts would appear to be an appropriate proxy for neighborhoods, for example, when there is an interest in investigating the possible "targeting" of neighborhoods. In addition, a longitudinal research design coupled with the use of quasi-experimental equivalent control groups is superior to cross-sectional designs in demonstrating whether market dynamics or discriminatory intent explains observed disparities. The search for definitive research designs in the assessment of inequity also raises difficult issues concerning longitudinal endogeneity and requires some understanding of the relevance of local zoning regulations in determining the location of new industrial firms.

Extending environmental justice research to include issues of risk and exposure raises new challenges, but also new possibilities to discern environmental inequities. The new risk-based emphasis may be more successful in producing evidence of inequities because risk and exposure are clearly linked to other arenas, such as public health and occupational safety, in which racial and ethnic inequalities are known to exist. While research based on the siting hypothesis must work toward incorporating more sophisticated models of specification that make use of a longitudinal design and appropriate control group comparisons, research based on the disparate risk hypothesis faces other methodological challenges.

First, researchers have to take into account that it is difficult to demonstrate a causal relationship between exposure to noxious sites and health effects, unless there is a particularly strong association or the population under study is large. Both conditions are not usually given in environmental justice–oriented studies. Second, there is the challenge of methodologically controlling for the fact that many populations will be exposed to multiple risk factors. These factors not only make them more susceptible to potential health risks associated with proximity to noxious sites, but also make it difficult to demonstrate which factors or combination of factors, including those with synergistic effects, are responsible for observed health outcomes. Environmental justice research will have to move beyond a single substance methodology to capture the effect of synergies between chemical substances. The issue of cumulative burden must also be dealt with, since some populations will be exposed to the same substance in multiple settings or will be exposed to substances with synergistic effects. In adopting a wider lens, environmental justice research would benefit from considering other issues pertinent to the spatial configuration of health beyond environmental burdens. Particularly relevant would be distributive issues pertaining to preventive health (e.g., the distribution of bike paths), since these may be related to the greater susceptibility of certain populations. By considering the larger urban environment nexus, environmental justice research has the potential to adopt a more integrated perspective. An integrated perspective is necessary, since the reliance on a purely empirical risk-oriented framework with increasingly more rigorous calculations may encounter problems in proving that exposure presents serious health risks. Meeting these challenges is crucial to the study of environmental inequities in siting and exposure.

NOTES

1. Evan J. Ringquist, "Equity and the Distribution of Environmental Risk: The Case of TRI Facilities," *Social Science Quarterly* 78, no. 4 (December 1997): 811–29; E. F. Pellizzari, R. I. Perritt, and C. A. Clayton, "National Human Exposure Assessment Survey (NHEXAS): Exploratory Survey of Exposure among Population Subgroups in EPA Region V," *Journal of Exposure Analysis and Environmental Epidemiology* 9 (1999): 49–55.

2. John A. Hird, "Environmental Policy and Equity: The Case of Superfund," *Journal of Policy Analysis and Management* 12, no. 2 (1993): 323–43.

3. Vicki Been, "Locally Undesirable Land Uses in Minority Neighborhoods: Disproportionate Siting or Market Dynamics?" *Yale Law Journal* 103, no. 6 (1994): 1383–1422.

4. Robert D. Bullard and Beverly H. Wright, "Environmental Justice for All: Community Perspectives on Health and Research Needs," *Toxicology and Industrial Health* 9, no. 5 (1993): 821–41.

5. Robert D. Bullard, "Environmental Racism and 'Invisible' Communities," *West Virginia Law Review* 96 (1994): 1037–50.

6. Eric Sheppard, Helga Leitner, Robert B. McMaster, and Hongguo Tian, "GIS-Based Measures of Environmental Equity: Exploring Their Sensitivity and Significance," *Journal of Exposure Analysis and Environmental Epidemiology* 9 (1999): 18–28.

7. For an example of a study based on zip code areas, see United Church of Christ (UCC) Commission for Racial Justice, *Toxic Wastes and Race in the United States: A National Report on the Racial and Socio-Economic Characteristics of Communities with Hazardous Waste Sites* (New York: United Church of Christ, 1987). A study using county-level data is Hird, "Environmental Policy." Two U. Mass. studies using tract-level data include Andy B. Anderson, Douglas L. Anderton, and John Michael Oakes, "Environmental Equity: Evaluating TSDF Siting over the Past Two Decades," *Waste Age* 25, no. 7 (July 1994): 83–100; Pamela Davidson and Douglas L. Anderton, "Demographics of Dumping II: A National Environmental Equity Survey and the Distribution of Hazardous Materials Handlers," *Demography* 37, no. 4 (2000): 461–66. An example of an environmental justice study using block-level data is provided by Janet T. Heitgerd, Jeanne R. Burg, and Henry G. Strickland, "A Geographic Information Systems Approach to Estimating and Assessing National Priorities List Site Demographics: Racial and Hispanic Origin Composition," *International Journal of Occupational Medicine and Toxicology* 4, no. 3 (1995): 343–63.

8. Susan L. Cutter, Danika Holm, and Lloyd Clark, "The Role of Geographic Scale in Monitoring Environmental Justice," *Risk Analysis: An Official Publication of the Society for Risk Analysis* 16, no. 4 (1996): 519.

9. Douglas L. Anderton, Andy B. Anderson, John Michael Oakes, and Michael R. Fraser, "Environmental Equity: The Demographics of Dumping," *Demography* 31, no. 2 (1994): 229–48.

10. See, for example, Vicki Been and Francis Gupta, "Coming to the Nuisance or Going to the Barrios? A Longitudinal Analysis of Environmental Justice Claims," *Ecology Law Quarterly* 24, no. 1 (1997): 1–56; Douglas L. Anderton, "Methodological Issues in the Spatiotemporal Analysis of Environmental Equity," *Social Science Quarterly* 77, no. 3 (1996): 508–15; Theodore S. Glickman, "Measuring Environmental Equity with Geographical Information Systems," *Resources* (Summer 1994): 2–6; Cutter et al., "Role of Geographic Scale."

11. Eric J. Krieg, "A Socio-Historical Interpretation of Toxic Waste Sites: The Case of Greater Boston," *American Journal of Economics and Sociology* 54, no. 1 (1995): 1–14.

12. Douglas S. Massey and Nancy A. Denton, "Residential Segregation of Blacks, Hispanics, and Asians by Socioeconomic Status and Generation," *Social Science Quarterly* 69, no. 4 (December 1988): 797–817.

13. Cutter et al., "Role of Geographic Scale."

14. Paul Mohai, "The Demographics of Dumping Revisited: Examining the Impact of Alternate Methodologies in Environmental Justice Research," *Virginia Environmental Law Journal* 14 (1995): 615–53.

15. Mohai, "Demographics of Dumping."

16. Been and Gupta, "Coming to the Nuisance."

17. Bullard, "Environmental Racism."

18. Environmental Protection Agency, *Final Guidance for Incorporating Environmental Justice Concerns in EPA's NEPA Compliance Analysis* (Washington, D.C.: Office of Federal Activities, April 1998).

19. John J. Fahsbender, "An Analytical Approach to Defining the Affected Neighborhood in the Environmental Justice Context," *N.Y.U. Environmental Law Journal* 5 (1996): 120–80.

20. Mohai, "Demographics of Dumping."

21. Allen C. Goodman, "A Note on Neighborhood Size and the Measurement of Segregation Indices," *Journal of Regional Science* 25 (1985): 471–76.

22. Cutter et al., "Role of Geographic Scale in Monitoring Environmental Justice," *Risk Analysis: An Official Publication of the Society for Risk Analysis* 16, no. 4 (1996): 525.

23. See Anderton, "Methodological Issues"; John Michael Oakes, Douglas L. Anderton, and Andy B. Anderson, "A Longitudinal Analysis of Environmental Equity in Communities with Hazardous Waste Facilities," *Social Science Research* 25 (1996): 125–48; Been, "Locally Undesirable Land Uses"; Been and Gupta, "Coming to the Nuisance."

24. Vicki Been, "Analyzing Evidence of Environmental Justice," *Journal of Land Use and Environmental Law* 11, no. 1 (1995): 1–36; Mohai, "Demographics of Dumping."

25. Been, "Analyzing Evidence"; Mohai, "Demographics of Dumping."

26. Been and Gupta, "Coming to the Nuisance"; Oakes et al., "Longitudinal Analysis."

27. Anderton, "Methodological Issues."

28. Oakes et al., "Longitudinal Analysis." As best described by Been and Gupta, "Coming to the Nuisance," the market dynamics argument posits that minorities and low-income residents "come to the nuisance" because of lower property values that make housing more affordable.

29. Charles P. Lord and William A. Shutkin, "Environmental Justice and the Use of History," *Environmental Affairs* 22 (1994): 1–26.

30. Lord and Shutkin, "Environmental Justice," 5.

31. Been and Gupta, "Coming to the Nuisance"; Oakes et al., "Longitudinal Analysis."

32. Oakes et al., "Longitudinal Analysis."

33. See, for example, General Accounting Office (GAO), *Siting of Hazardous Waste Landfills and Their Correlation with Racial and Economic Status of Surrounding Communities* (GAO/RCED-83-168) (Washington, D.C.: General Accounting Office, 1983); UCC Commission for Racial Justice, *Toxic Wastes and Race;* Paul Mohai and Bunyan Bryant, "Environmental Racism: Reviewing the Evidence," in *Race and the Incidence of Environmental Hazards: A Time for Discourse,* ed. Bunyan Bryant and Paul Mohai (Boulder, Colo.: Westview, 1992).

34. Davidson and Anderton, "Demographics of Dumping."

35. Bullard, "Environmental Racism."

36. Anderson et al., "Environmental Equity."

37. Been and Gupta, "Coming to the Nuisance."

38. The Council on Environmental Quality issued a draft guidance already in May 1996 to help federal agencies to incorporate environmental justice concerns in their NEPA procedure (for a summary, see Ronald Bass, "Evaluating Environmental Justice under the National Environmental Policy Act," *Environment Impact Assessment Review* 18, no. 1 [1998]: 83–92). The final draft was released in April 1998. Even before this, President Clinton issued a memorandum accompanying his executive order in which he urged agencies to use the NEPA EIS requirement to achieve environmental justice goals (reported in Been and Gupta, "Coming to the Nuisance"). The *Final*

Guidance of 1998 formalized the president's encouragement and provided detail that is more explicit as to how these recommendations might best be met.

39. Carl V. Phillips and Ken Sexton, "Science and Policy Implications of Defining Environmental Justice," *Journal of Exposure Analysis and Environmental Epidemiology* 9 (1999): 9–17.

40. Phil Brown, "Popular Epidemiology: Community Response to Toxic Waste–Induced Disease in Woburn, Massachusetts," *Science, Technology, and Human Values* (Summer/Fall 1987): 78–85.

41. Bunyan Bryant, "Pollution Prevention and Participatory Research as a Methodology for Environmental Justice," *Virginia Environmental Law Journal* 14, no. 4 (1995): 589–613.

42. Bullard, "Environmental Racism."

43. Lance A. Waller, Thomas A. Louis, and Bradley P. Carlin, "Environmental Justice and Statistical Summaries of Differences in Exposure Distributions," *Journal of Exposure Analysis and Environmental Epidemiology* 9 (1999): 56–65.

44. Ringquist, "Equity," 114.

45. Sheppard et al., "GIS-Based Measures." The distance decay to which the authors refer describes the difficult-to-measure rate of decrease in the toxicity and potential hazardousness of chemicals with increasing distance from the site of release.

46. Siting biases along racial, ethnic, and class lines are thought to be the result of certain communities being targeted because they lack the appropriate information and political clout to prevent a siting of a facility in their community or the requisite economic resources to "vote with their feet" by moving out of host neighborhoods.

47. Mohai, "Demographics of Dumping."

48. W. M. Bowen, M. J. Salling, K. E. Haynes, and E. J. Cyran, "Toward Environmental Equity in Ohio and Cleveland," *Annals of the Association of American Geographers* 85, no. 4 (1995): 641–63; Michael S. Scott and Susan L. Cutter, "Using Relative Risk Indicators to Disclose Toxic Hazard Information to Communities," *Cartography and Geographic Information Systems* 24, no. 3 (1997): 158–71.

49. Glickman, "Measuring Environmental Equity."

50. Bullard, "Environmental Racism."

51. Brian D. Israel, "An Environmental Justice Critique of Risk Assessment," *New York University Law Journal* 3, no. 2 (1994): 496.

52. Ken Sexton and John L. Adgate, "Looking at Environmental Justice from an Environmental Health Perspective," *Journal of Exposure Analysis and Environmental Epidemiology* 9 (1999): 3–8.

53. S. A. Geschwind et al., "Risk of Congenital Malformations Associated with Proximity to Hazardous Waste Sites," *American Journal of Epidemiology* 135, no. 11 (1992): 1197–1207.

54. Scott and Cutter, "Using Relative Risk Indicators."

55. Data on TSDF sites may come from various sources. The data that I have used in empirical analysis come from the 1992 Environmental Services Directory (ESD), published by Environmental Information Ltd. which, after extensive cleaning (see Anderson et al., "Environmental Equity") contained 520 TSDFs. Been and Gupta's study of 608 TSDFs was based on lists from several sources, including RCRIS, the 1994 ESD, and several lists provided by trade organizations and other researchers (see Been and Gupta, "Coming to the Nuisance"). The 415 TSDFs analyzed in the

1987 UCC study were extracted from the 1986 Hazardous Waste Data Management System maintained by the EPA and the 1986 list of "Industrial and Hazardous Waste Management Firms" published by Environmental Services Ltd. (see UCC Commission for Racial Justice, *Toxic Wastes and Race*).

56. Shelley A. Hearn, "Tracking Toxics: Chemical Use and the Public's 'Right-to-Know,'" *Environment* 38, no. 6 (July/August 1996): 5–9, 28-30.

57. Pellizzari et al., "National Human Exposure Assessment."

III

RESPONSES TO ENVIRONMENTAL INJUSTICES

6

Syndrome Behavior and the Politics of Environmental Justice

Harvey L. White

Environmental justice issues have been prolific sources of political activity for several decades. Much of this activity is in response to decision-making processes surrounding environmental hazards. The response is so intense that it generates sets of syndrome behavior endemic to this body of politics. This chapter contains a discussion of these syndrome behaviors and the politics of environmental justice that they engender. It also contains case studies to illustrate these syndrome behaviors in the context of the American political system. Syndrome behaviors associated with environmental justice represent action taken when citizens are affected, or think they are going to be affected, by environmental hazards and actions by politicians who are forced to respond to citizens' concerns. These actions have spawned a new social movement; have received major attention from legislative, administrative, and judicial officials; and have been the impetus for a new body of research.

RESEARCH ON ENVIRONMENTAL JUSTICE

Numerous research studies have focused on environmental justice issues. Findings from these studies indicate that there are significant disparities in the distribution of risks from environmental hazards (table 6.1). Research suggests that people of color and low-income individuals are frequently and often severely exposed to potentially deadly and destructive levels of toxins from environmental hazards. These research findings constitute convincing evidence that this pattern of exposure to environmental hazards transcends almost every aspect of their lives; this includes places where they work, live, play, and learn, and the foods they eat. Racial disparities were found in

Table 6.1. Selected Studies of Racial and Income Disparities in Distribution of Environmental Hazards, 1977–1998

Year	Author	Type of Hazard	Geographic Focus	Disparity Race	Disparity Income
'77	Berry et al.	Pollution/pesticides, etc.	Urban Areas	Yes	Yes
'77	Kutz et al.	Pesticides	National	Yes	
'78	Asch & Seneca	Air Pollution	Urban Areas	Yes	Yes
'80	SRI	Toxic Fish	National	Yes	No
'81	Puffer	Toxic Fish	Los Angeles, CA	Yes	
'83	U.S. GAO	Hazardous Waste	Southeast	Yes	
'84	Greenberg & Anders.	Hazardous Waste	New Jersey	Yes	Yes
'85	McAllum	Toxic Fish	Puget Sound, WA	Yes	
'85	NOAA	Toxic Fish	Puget Sound, WA	Yes	
'86	Gould	Hazardous Waste	National	Yes	
'87	UCC & PDA	Hazardous Waste	National	Yes	Yes
'87	Gelobter	Air Pollution	Urban Areas	Yes	Yes
'88	ATSDR	Lead	Urban Areas	Yes	Yes
'89	Belliveau et al.	Toxic Releases	Richmond, CA	Yes	Yes
'89	Pfaff	Air Pollution	Detroit, MI	Yes	
'90	Cater-pokras et al.	Lead	National	Yes	
'90	Bullard	Toxic Waste	Southeast	Yes	
'91	Brown	Toxic Releases	St. Louis, MO	Yes	
'91	Costner & Thornton	Hazardous Waste	National	Yes	Yes
'91	Kay	Toxic Releases	Los Angeles, CA	Yes	
'91	Mann	Air Pollution	Los Angeles, CA	Yes	
'91	Wernette & Nieves	Air Pollution	Urban Areas	Yes	
'91	Goldman	Toxins	National	Yes	Yes

Year	Author	Topic	Location		
'92	Fitton	Hazardous Waste	National	Yes	Yes
'92	Goldman	Toxic Air/Waste	National	Yes	No
'92	Holtzman	Waste Incineration	New York, NY	Yes	
'92	Ketkar	Hazardous Waste	New Jersey	Yes	
'92	McDermott	Hazardous Waste	National	Yes	Yes
'92	Mohai & Bryant	Toxic Waste/Pollution	Detroit, MI	Yes	Yes
'92	Nieves	Hazardous Waste	National	Yes	
'92	Roberts	Hazardous Waste	New York, NY	Yes	Yes
'92	Unger et al.	Toxic Fish	Pinewood, SC	Yes	No
'92	West et al.	Hazardous Waste Siting	Michigan	Yes	Yes
'93	Been	Post Siting of Hazards	Southeast	Yes	No
'93	Burke	Toxic Releases	Los Angeles, CA	Yes	Yes
'93	Bowen et al.	Toxic Releases	Cuyahoga, OH	No	Yes
'93	Greenberg	Toxic Releases	Ohio	Yes	No
'93	Hamilton	Incinerators (Large)	National	Yes	Yes
'93	Zimmerman	Hazardous Waste Siting	National	Yes	No
'95	West	Hazardous Waste	Michigan	Yes	No
'98	Timney	Toxic Fish	Ohio	Yes	Yes
'98	Clarke & Gerlak	Toxic Releases	Arizona	Yes	Yes
'98	Wright	Toxins	Louisiana	Yes	Yes
		Toxic Releases			

Source: Benjamin Goldman, *Not Just Prosperity: Achieving Sustainability with Environmental Justice* (Washington, D.C.: National Wildlife Federation, 1994) and David E. Camacho, *Environmental Injustices, Political Struggles: Race, Class, and the Environment* (Durham, N.C.: Duke University Press, 1998).
Terms: NOAA (National Oceanic and Atmospheric Administration)
SRI (Superfund Redevelopment Initiative)
PDA (Public Data Access Inc.)

87 percent of the studies, and income disparities were found in 74 percent. Disparities were found to exist in a variety of areas (i.e., exposure to toxins, siting of hazardous facilities, solid waste, and occupational health). These disparities were also observed in all regions of the country and in both urban and rural communities. Table 6.1 summarizes these studies. Scholars who conducted the studies are from thirteen different professional fields and used a variety of research methods. Not only do their findings constitute a persuasive body of evidence, but they also suggest the need for concern about the possible effects of environmental hazards on people of color and low-income individuals.

Why the Concern about Environmental Hazards?

All communities are affected by environmental hazards, including European Americans, African Americans, Native Americans, Latino Americans, and Asian Americans. Arguably, these hazards represent one of society's greatest health risks. In several respects, environmental hazards pose a greater health threat than the dreaded AIDS virus. In the case of AIDS, there is something that individuals can do to protect themselves. They can restrict their exposure to HIV by abstaining from risky sexual, drug-related, and other activities. In other words, they can isolate themselves almost completely from media through which the virus is transmitted. If they choose to expose themselves to media through which the virus is transmitted, there are protective measures readily available that can be employed to minimize the risk of contracting AIDS. In contrast, when environmental hazards exist, there is virtually nothing that individuals can do to protect themselves. First, and most crucial, they cannot limit their exposure to the media that transmit environmental hazards, which are the air, food, and water necessary for survival. Second, there are no measures that are readily available, that can easily be employed, to minimize the risk of coming in contact with environmental hazards. Once environmental hazards exist, the very life-sustaining functions that individuals must perform may put them at risk of exposure to life-threatening toxins. Thus, it is almost impossible to protect oneself from environmental hazards, because individuals have virtually no control over the quality of the air they breathe, the food they eat, or the water they drink. They are almost completely dependent upon someone else to protect them from environmental hazards. Concern about this dependence contributes to the intense activity generated in response to environmental hazards. Decisions associated with locating and siting environmental hazards have been the focus of most of this activity. These decisions have affected all sectors of society. No events in recent history have drawn a greater cross section of the American public into active politics. Neither rich nor poor, black nor white, young nor old have remained detached from the politics of these decisions.

POLITICS AND "THE POLITICS" OF ENVIRONMENTAL JUSTICE

There are many perspectives on what constitutes politics. One is that it has to do with the process through which authoritative decisions are made binding. Citizens are said to view such decisions as binding because of habit, tradition, respect for certain procedures, loyalty to persons or institutions, and fear. Activities emanating from these processes are believed to be political when they are directed toward defining these decisions, determining who the decision makers shall be, implementing the decisions, and resolving any conflicts over their meaning or application. According to David Camacho, "Politics refers to that set of social structures and processes by which humans resolve their conflicting interests without having to be in a constant Hobbesian 'state of war'."[1] At its most elementary level, as Harold Lasswell observed more than thirty years ago, politics is the set of intense processes that determine "who gets what, when, and how."[2]

The politics of environmental justice is no less intense and perhaps in some ways more so. This politics entails a set of activities directed toward influencing decisions concerning environmental hazards. Or, to paraphrase Lasswell, the politics of environmental justice is a response to processes that decide:

Who gets environmental hazards dumped on them;
What environmental hazards get dumped on them;
When environmental hazards get dumped on them;
Where environmental hazards get dumped on them;
How individuals and communities respond when environmental hazards are dumped on them, or how they respond when they think environmental hazards are going to be dumped on them; and
How they cause others to respond to them when environmental hazards are dumped on them or think environmental hazards are going to be dumped on them.[3]

Responses to these processes are so intense that they constitute what might be classified as syndrome behaviors.[4]

SYNDROME BEHAVIORS AND ENVIRONMENTAL HAZARDS

Several syndrome behaviors have been identified that are often associated with decision making surrounding environmental hazards. These include:

NIMBY	(Not In My Back Yard)
NIMTOO	(Not In My Term Of Office)
NIMEY	(Not In My Election Year)

PIITBY   (Put It In Their Back Yard)
WIMBY   (Why In My Back Yard?)

These syndromes have caused politicians to wither in the face of their constituents. They have caused costly delays for organizations seeking to develop waste treatment or related facilities. Syndrome behaviors associated
with environmental hazards do not occur in a vacuum. They are "social
products" of the dynamic circumstances that develop as individuals react
and interact in response to environmentally sensitive situations.

NIMBY BEHAVIOR–POWER POLITICS

"Not-In-My-Back-Yard" (NIMBY) is one of the responses to environmental
hazards. Given a choice, no community is likely to want a waste facility in its
backyard. Any mention of such facilities usually results in intense opposition. However, not every community has the resources to practice NIMBY
political behavior. The NIMBY syndrome is usually observed in more economically and politically affluent communities. NIMBY political behavior is
proactive. As William Glaberson of the *New York Times* points out, NIMBYs
organize, march, sue, and petition to block the developers they think are
threatening them. They use the political and legal systems to cause inter-
minable delays. Potential cost overruns associated with these delays often
cause projects related to environmental hazards to be canceled.[5]

Residents in more affluent communities have the time, money, and knowledge, which are crucial resources for any NIMBY campaign, to practice
NIMBY politics. Recognizing the negative as well as the positive implication
of science and technology is a necessary first step. Time and monetary resources are needed only after this recognition has occurred. When these
three resources are available to a community, the NIMBY political behavior
is likely to be the response, if the community feels threatened by an environmental hazard. Many examples of NIMBY behavior have been cited in
the literature. An occurrence in western Pennsylvania also typifies this behavior. Residents in Clarion County, Pennsylvania, have exhibited classical
NIMBY behavior in response to information that their community was being
considered for a proposed hazardous waste facility.

NIMBY in Clarion County, Pennsylvania

According to Walter Rosenbaum, "NIMBYism is rarely routed by better information, more qualified experts, improved risk communication techniques, and other palliative actions."[6] Its ranks crowded with well-educated,
socially active, and organizationally experienced individuals, NIMBYism in

Clarion, Pennsylvania, is not only a tough, stubborn, and durable opponent of those who would locate a hazardous waste facility in this community, but it is also a formidable challenge to environmental officials who make permit and regulatory decisions. Clarion County is a college community of 41,000. It has a nearly homogeneous population. Nearly all (99.97 percent) of the residents are European Americans. Only 12 percent of the population is over sixty-five. The county's economy is based on a mixture of agriculture, manufacturing, construction, and service industries. The average family income is more than $15,000 a year. The residents of Clarion are well educated. Nearly 75 percent have between twelve and sixteen years of schooling. The county is also home to Clarion University. By most standards, Clarion is a viable community. In normal times, Clarion is a quiet rural community. Feature events include high school football games and the autumn leaf festival. However, things changed when the Concord Resources Group announced that the county was being considered for its hazardous waste site.

"The Study"

The expeditious and integrated manner in which residents responded to the announcement is astounding. The Concord Resource Group announced in August 1990 that it was studying two sites in Clarion County to determine if either was suitable for a $100 million hazardous waste complex, which would include a landfill, an incinerator, and treatment works. Typical of NIMBY behavior, a full-fledged political movement developed in Clarion before any major decision was made. Community meetings were organized to illustrate mass opposition to the facility. The announcement of the study resulted in the politics of NIMBYism in Clarion County within only a few days.

The Clarion political movement that developed in opposition to the proposed hazardous waste facility involved every segment of the community in the NIMBY campaign. Virtually every business in Clarion displayed a sign or a bumper sticker in its window. Students circulated petitions and wrote letters asking the governor to stop the facility. Many residents posted signs in their front yards that announced to officials from the Pennsylvania Department of Environmental Resources, "DER, We're Watching You." An official coordinating group (PEACE—Protect the Environment and Children Everywhere) was established to raise money and lobby public officials. At its first meeting, $1,000 was raised. Within a few days after the announcement that the county was being considered for a waste site, PEACE revealed plans to raise $250,000 to hire lawyers and scientists to fight the facility. Fund-raising activities began immediately. Within six weeks, more than 21,000 residents signed a petition opposing the facility. More than 8,000 joined PEACE. The building and supply store sold hats and T-shirts to help the group raise money.

Almost every elected official in the county was persuaded to oppose the facility. Clarion County hired an attorney and the same environmental consultants who helped Braintree, Massachusetts, win its four-year battle against a proposed toxic waste facility. The community formed a brain trust to match wits with the company and to monitor activities of state regulatory officials. This all-volunteer brain trust included archaeologists, historians, hydrologists, toxicologists, geologists, attorneys, librarians, university professors, teachers, and a host of retirees and housewives. These volunteers spent thousands of hours poring over aerial photographs, maps, gas well records, and numerous state and federal laws and regulations. A crucial aspect of their task was making sure that environmental managers understood and enforced state and federal laws. Concern in this area prompted one volunteer to observe, "All we have asked from the beginning is: Does the state know what it is doing?" Volunteers made state environmental officials aware of potential wetland destruction on one of the proposed sites. An active gas well was also discovered on the property. Each might be considered a reason for rejecting the permit application for the hazardous waste facility.

As one resident noted, "We know this is a psychological war. We're dealing with a corporation with big bucks that knows how to play the mind games. Well, guess what? We can play those games." Residents of Clarion have played exceptionally well. The application to proceed with development of the facility has been rejected twice by the Pennsylvania Department of Environmental Resources. NIMBYs in Clarion County were able to "organize, march, sue and petition" to block the hazardous facility threatening them. Hence, after numerous delays caused by various NIMBY activities, the Concord Resources Group eventually decided in April 1993 not to pursue development of the hazardous waste facility. As in other affluent communities, NIMBYism was able to prevail in Clarion County.

NIMTOO BEHAVIOR—ISSUES OF POLITICAL LEGACY

The intense and pervasive nature of NIMBY syndrome behavior has led to development of an evasive political behavior that can be characterized as the NIMTOO (Not In My Term Of Office) approach to decision making when environmental hazards are involved. Politicians have found themselves caught in the middle of the struggle between industries' concerns about inadequate disposal facilities for environmental hazards and citizens' concerns about the health threats from environmental hazards. Worried about how highly controversial decisions in these areas will affect their careers, they have frequently refused to pass laws and have delayed the implementation of existing regulations during their terms of office. This is a prevalent behavior at all levels of government and has major implications for addressing issues surrounding environmental hazards. Political scientists have largely ignored this behavior.

Typical of the NIMTOO approach has been political behavior associated with efforts to find permanent storage facilities for nuclear waste. Several successions of federal and state officials delayed the development and implementation of a national policy to eliminate the impending nuclear waste storage problem. Even though the storage problem had been a concern for nearly five decades, elected officials at every level avoided proposed solutions. It has generally been considered a technical problem by both the government and the nuclear industry. Most of the nuclear waste, which must be disposed of, is generated by privately owned nuclear facilities. The nuclear industry counted on the federal government's policy of reprocessing spent fuel to alleviate the waste problem. However, the government discontinued this policy during the Jimmy Carter administration. Shortly after this decision, a number of states enacted moratoriums on nuclear plant construction, based on the rationale that without adequate waste disposal facilities, nuclear energy is unpredictable and uneconomical. The courts upheld these moratoriums. The response has resulted in a series of NIMTOO-related behaviors, involving elected officials at all levels of government. All of these officials have acknowledged the need for a permanent solution for the waste problem, but not in their term of office.

Perhaps the most pronounced set of NIMTOO behaviors has been that centered on siting requirements of the Nuclear Waste Policy Act (NWPA; see table 6.2). The NWPA mandated a nationwide search for two permanent high-level waste disposal sites in a variety of geologic media. One site was to be in the West and the other in the East. After intense opposition from state and local officials, Congress amended the act to concentrate site evaluation at Yucca Mountain in Nevada. Evaluation of this location was expected to take approximately five years and, if it were not acceptable, Congress would reconsider the issue. Although the long delay in making the final decision was in part due to several delays caused by technical limitations, no group of officials seemed to want the distinction of having made siting decisions for nuclear waste. It took nearly twenty years for a conclusive decision to be made on sites for the storage of nuclear waste. This included congressional and administrative delays during the terms of four presidents. The technical limitations on the disposal of nuclear waste are quite challenging and should not be underestimated. Overcoming these limitations requires a tremendous commitment of time, money, and determination. They already have the focus of many engineers, physicists, chemists, and other physical scientists. However, the activities associated with the NIMTOO behavior described above do not require technical solutions. They require a better understanding of the social behavior associated with the politics of environmental justice.

Table 6.2.   Nuclear Waste Policy: A First Fifty Years Chronology

1945   The first nuclear weapons are produced by the United States.

1954   Peaceful use of atomic energy promoted with the federal government responsibility for disposal of radioactive waste.

1956   The National Academy of Sciences recommends deep geologic disposal of the long-lived, highly radioactive wastes.

1960s   Michigan and Ohio officials stop investigation of the salt beds of the Salina Basin as a nuclear waste disposal site.

1970s   Abandoned oil and gas exploration boreholes force reversal decision to use a salt mine, at Lyons, Kansas.

1982   Congress passes the Nuclear Waste Policy Act, requiring two repositories for disposal beginning in 1998.

1983   The Department of Energy names nine previously screened, potentially acceptable repository sites in six states.

1986   The DOE selects three western sites, in Nevada, Texas, and Washington, for detailed investigation.

1986   The second repository site screening program is postponed after objection from states in the northern Midwest and East.

1986   The DOE proposes an interim monitored retrievable storage (MRS) facility for commercial waste in Tennessee.

Late 1987   Congress names Yucca Mountain as the only site to be characterized for development as a repository.

1989   The secretary of energy determines a new program strategy for waste acceptance beginning at a repository in 2003.

1990   The National Academy of Sciences determines that DOE needs more flexibility for licensing a repository.

1992   DOE testifies that licensing regulations are causing delays and escalating costs in the Yucca Mountain Project.

1992   Congress instructs EPA to establish new site-specific environmental regulations for Yucca Mountain.

1992   Efforts of the nuclear waste negotiator to provide a volunteer MRS site fail.

1993–1994   A new Program Approach is developed (adopted in 1994) that sets the acceptance of waste in 2010.

1995   Congress puts highest priority on interim waste storage, at Yucca Mountain, in 1998, or as soon as possible.

1995   A new schedule is developed for the Yucca Mountain Project.

1996   DOE's new Program Plan is completed. Congressmen develop interim storage site plan for Nevada Test Site in 1998.

1996   Congress directs that a viability assessment of the Yucca Mountain project be delivered by October 1998.

1996 DOE issues a notice of proposed rulemaking to revise site suitability guidelines for a Yucca Mountain repository.

1997 Bills emphasizing interim storage of spent fuel at Nevada Test Site are introduced. President Clinton says he will veto bill.

Jan. 1998 DOE issues a revision of its Waste Isolation and Containment Strategy.

Dec. 1998 DOE issues the viability assessment for a Yucca Mountain repository.

Feb. 1999 NRC proposes a new repository licensing rule, specific to licensing a Yucca Mountain repository.

1999 Bills emphasizing interim storage at Nevada Test Site are introduced again. President Clinton says he will veto bill.

NIMEY BEHAVIOR—ELECTION POLITICS

Like NIMTOO behavior described above, the "Not In My Election Year" (NIMEY) syndrome also emanates from motivation associated with political ambition. Unlike the NIMTOO emphasis on how one's term in office will be perceived, NIMEY behavior is focused on being reelected to office. As noted earlier, politicians are motivated by the desire to survive and advance politically. "Political survival and advancement are tied to constituency service and local interest, which in turn are tied to . . . reelection-survival in the basic sense."[7]

Decisions about nuclear waste disposal are also illustrative of NIMEY behavior and the dilemma of electoral politics. For instance, it has not been unusual for waste siting decisions with electoral implications to be delayed or canceled. Hearings are held, studies are conducted, and legislation enacted, but formal decisions are withheld until after crucial elections. As Gerald Jacob observed in his discussion of the politics of siting a nuclear waste depository, "The announcement that Texas held one of three sites chosen for characterization was delayed until after the 1984 elections."[8] NIMEY behavior was displayed again in 1987 (before the 1988 elections), when work at all waste sites, except Yucca Mountain, was halted; no consideration was given to the selection of a second repository in the eastern United States.[9] Nicholas Lenssen describes a series of delays associated with the NIMEY behavior in his discussion of the nuclear waste problem. "In 1975, the United States planned on having a high-level waste burial site operating by 1985. The date was moved to 1989, then to 1998, 2003, and now 2010."[10]

NIMEY syndrome behavior is also typified by what is referred to as "the tyranny of the immediate." That is, politicians tend to be preoccupied with the short-run consequences of elections. "A desire to sidestep possible intractable legislative conflicts by shifting them to administrators, or to avoid the threatening political and economic consequences of enforcing laws, is instinctive to all politicians."[11] Although it is difficult to generalize, politicians are often described by scholars in the field as ambitious risk takers who are willing to live frantic lifestyles. They usually have to be balancers and bargainers who try to reconcile competing claims of what is the "public interest." Perhaps most importantly, they "have to follow public opinion as well as shape or mold it."[12] It is the latter quality that gives rise to NIMEY behaviors. That is, the nature of environmental issues makes it almost impossible to follow or shape public opinion. As Carol Barner-Barry and Robert Rosenwein explain in *Psychological Perspectives on Politics,* such "dilemmas of political leadership" occur when there are changes in macroenvironmental factors or when the perceptions and expectations of the electorate have changed.[13] Accordingly, the behavior of politicians must change. The NIMEY syndrome is one of the political behaviors that has become endemic to politics that surround environmental justice.

PIITBY BEHAVIOR—AN EXAMPLE OF POLITICAL COMPROMISE

Political compromise is an established tradition in American politics. It is an art that most politicians learn early in their careers. Compromise is thought to be rational political behavior in which two opposing spheres of power have vastly different opinions on matters of mutual interest. It is practiced in every political arena.

A Zero-Sum Game

The compromise achieved is often at the expense of those who are not represented when negotiations are being held. The politics of environmental justice is no exception. Pressure for a solution to problems associated with environmental hazards frequently forces officials to look for a compromise. The likely compromise is to "Put It In Their Back Yard" (PIITBY).

The PIITBY compromise often results in a decision to place environmental hazards in minority, low-income, or politically weak communities. Circumstances internal and external to these communities encourage their selection as sites for the location of environmental hazards. The priority exhibited in site selection is one such circumstance. Principally, sites given the most attention will be those that affect the more economically or politically affluent communities. Such communities will have the resources, knowledge, and contacts to sustain the symptoms of the NIMBY syndrome. As a consequence, residents from these communities are more likely to be proactive. They are, generally, the driving force that causes politicians to exhibit both the NIMBY and NIMTOO syndromes.

WIMBY—THE "WHY IN MY BACK YARD" POLITICAL BEHAVIOR

The NIMBY, NIMEY, NIMTOO, and PIITBY syndromes are likely to be exhibited less frequently in low-income, minority, and other communities that are not affluent. These communities are more prone to exhibit a WIMBY ("Why In My Back Yard") syndrome. That is, they are more reactive than proactive in their responses to decisions concerning environmental hazards. This WIMBY syndrome emanates from social, economic, and political realities that surround these communities. These less-affluent communities usually do not have the resources or contacts to initiate or sustain the proactive behavior necessary for successful NIMBY behavior. Nor do residents in these communities have the contacts in government and industry necessary to become involved during preplanning and planning stages that precede crucial environmental policy and management decisions. These factors and others have led to a "knowledge and information" gap in low-income and minority communities about environmental hazards. Perhaps, because of the tradition of having landfills and other waste facilities in their communities, there is also more of a "social acceptance" of environmental management decisions.[14] Thus, the activism or WIMBY syndrome prevalent in less-affluent communities tends to develop after facilities have been constructed or other crucial decisions have been made.

The WIMBY syndrome is also far more congenial to the "Not In My Election Year" and the "Not In My Term Of Office" political behaviors than is the NIMBY syndrome. For politicians, it is safer to investigate why something was done than to intervene while something is being done. The "why" is less likely to affect voter decisions. Most of the politically sensitive decisions will have already been made when symptoms of the WIMBY syndrome become apparent. Decisions about zoning, building permits, and franchise licenses can occur with little or no public outcry.

The greatest concern in low-income and minority communities is raised after facilities are operational. It is then that residents learn of the serious threat that can be posed by environmental hazards. The consequence of siting Rollins's hazardous incinerator in Alsen, Louisiana, is a "clear case in point."

The "Unfriendly Neighbor"

Although the Rollins facility in Alsen evolved from a landfill to its current status as a hazardous waste incinerator, it was viewed as an "unfriendly neighbor" shortly after arriving in the community. As noted by one Alsen resident,

> When Rollins purchased that land, they pursued their operations as an unfriendly neighbor because they started digging all over the place. Large holes

were dug. . . . Large trucks with trailers began entering and exiting this facility with all types of materials and waste disposal. People who worked there from the Alsen Community were not allowed to touch the trailers. . . . Odorous vapors used to be seen and smelt throughout the community. This brought about the identification of the business Rollins was conducting.[15]

Rollins's initial waste disposal facility in Alsen was established in 1971. It was developed as a landfill disposal unit for hazardous materials. By 1981, this facility included a hazardous waste incinerator. Landfills at the Alsen site proved to be problematic for Rollins. Contaminated surface water from landfill sites ran into the river and was absorbed by the soil, causing a "topographical downgrade." As a result of a consent order with EPA, monitor wells were installed from December 15, 1980, until February 1981 and a site remediation program was developed. Rollins subsequently installed an incineration method to dispose of PCBs. According to company informational materials, the incinerator disposes of hundreds of pounds of hazardous waste daily.

Although Alsen residents considered operations at the landfill to be annoying, it was activities at the incinerator that led residents to exhibit the WIMBY syndrome behavior. These activities have antagonized, frightened, and frustrated local residents. They have also resulted in lawsuits and efforts to close the facility. WIMBY behavior has not been successful in removing the hazardous waste facility from the Alsen community.

WIMBY in Alsen, Louisiana

The Alsen Community is located in central Louisiana, ten miles north of the state capital (Baton Rouge). It is strategically located with geographical boundaries on the Mississippi River and along two major north-south highways. It is serviced by major rail lines and is only four miles from the Baton Rouge Metropolitan Airport. Nearly all of the residents in Alsen (98 percent) are African American. Based on social and economic indicators, Alsen should be a viable middle class community with a promising future. The average income is above $15,000 a year. Nearly 80 percent of the population have at least some college training. The community is also the former home of a historically black college. More than half (50 percent) of the residents have lived in the community for ten years or more. Only 9 percent are over sixty-five, with the others distributed fairly evenly among various age groups. These statistics suggest that Alsen is a wholesome place to live. Indeed, residents in this semirural community had few concerns until Rollins Environmental Services located its waste facility in the neighborhood.

Some Consequences of Hazardous Waste Incineration in Alsen

The responses to activities at the Rollins incinerator in Alsen illustrate behaviors that emanate from the WIMBY syndrome. Even though the air has been polluted, church services have been disrupted, and the health of preschoolers has been threatened, Alsen residents have been unable to have the facility removed from their community. The quiet, serene, and wholesome pre-Rollins community no longer exists. Residents complained for a number of years of chronic air pollution and other environmental problems, to no avail. They reported contaminated water, ill-smelling odors, eye irritations, and vegetation and property damage. It took a series of near disasters before public health officials began to investigate the consequences of activities at the hazardous waste facility.

On August 5, 1985, the incinerator at the Rollins hazardous waste facility malfunctioned, sending heavy black smoke and odors throughout the community. Residents reported feeling nausea, eye irritations, and other afflictions. The religious service at the Mount Bethel Baptist Church was terminated because of pollutants. According to Reverend W. L. Fontenot, he had to cut his sermon short during a revival meeting that night because he had a headache and his eyes were burning. Others at the church also experienced similar symptoms and discomforts. A similar incident on February 6, 1986, threatened the health of students and staff at a Head Start center. According to a report in the *New Orleans Times-Picayune:*

> [F]umes from a hazardous waste company apparently made pre-schoolers at a Head Start center sick. . . . Rollins failed to report last week's incident to the State as required by State regulations. . . . Classes at the Alsen Head Start Center were disturbed Thursday by a foul odor that caused some of the 3- and 4-year olds to vomit. . . . The problem occurred about the same time as a release of fumes at the plant during a waste mixing operation.[16]

In addition to the two incidents mentioned above, the hazardous waste facility has been cited more than one hundred times for violations. Such violations led to a state fine of $1.7 million. Health threats, as a consequence of activities at the facility, were the subsequent concern of state officials. George Cramer, an administrator with the Louisiana Department of Environmental Quality, noted, "It's been 2½ years since the company got the first test results (showing contamination), and Rollins did nothing to rectify the situation. We are concerned that if it is not taken care of it will represent a risk."[17] State officials had come to realize what Alsen residents had known for some time: The Rollins facility represented a significant health risk and had reduced the quality of life in the community. A research study reported high concentrations of sulfur dioxide, nitrogen oxides, carbon monoxide, hydrocarbons, and acids in areas surrounding the facility. Extensive exposure to any of these can lead to severe health problems.

The consequences of activity at the Rollins facility forced the community to seek relief in the courts. By 1986, nine lawsuits had been filed against Rollins by Alsen residents. The most renowned is *McCastle II. McCastle II* is reported to have been settled with an award of $350,000 to Alsen residents. Although the community has not been hesitant to take legal action against Rollins, it has not been able to prevent the incinerator from emitting toxic pollutants.

When the pollutants described above were released, Rollins was under an injunction barring it "from emitting chemical fumes and odors that make the plaintiffs {McCastle et al.} ill or cause serious discomfort." Rollins took a more aggressive posture after complaints were made about releases from its facilities. After the February 6, 1986, incident, Rollins responded "by dispatching a lawyer, its local public relations official and a stenographer to the [day care] center to take depositions from parents and teachers."[18] Although Rollins defended this action by saying that "it was trying to determine the facts and be a good neighbor," residents suggested that this was just another intimidation tactic by the company. According to Alsen residents, the neighborly thing for Rollins to do is to close the hazardous waste facility.

Rollins has conceded that the plant does not have a model record. However, even with its record of violations, Rollins has been able to resist efforts to close the facility. It was even able to keep the facility open when the Louisiana secretary of environmental quality ordered it closed after equipment malfunctioned. The company quickly countered the order to close with legal petitions and was able to get court permission to reopen the plant over the secretary's objection. It was also able to get the state supreme court to bar the secretary from serving as the hearing officer to determine the fate of the hazardous waste incinerator. Although the facility has a record replete with violations and unneighborly activity, it still operates its hazardous waste incinerator in the Alsen community. To say the least, the facility has had a major impact on residents and their perceptions of themselves and the community.

Residents' behavior in response to the incinerator was reactive and thus exhibited symptoms associated with the WIMBY syndrome. However, the activities at the hazardous waste incinerator have also led to proactive behavior. Alsen residents seem to have narrowed the "knowledge and information gap." They have become astute participants in the public policy process in Baton Rouge. Residents of Alsen are networking with environmental agencies and other communities. They are focusing on state and national environmental issues. They have successfully challenged proposals to place additional hazardous waste facilities in their community. One can only wonder what might have happened if this proactive behavior had been there to challenge the initial decision to place Rollins's hazardous waste incinerator in Alsen.

TWO VIEWS OF THE POLITICS OF ENVIRONMENTAL JUSTICE

As illustrated by the Alsen and Clarion County cases, the politics of environmental justice can lead to both productive and tragic outcomes. Consequently, these politics can be seen from positive and negative points of view.

From the negative perspective, NIMBYism can be described as a politics of division. Similarly, the WIMBY politics might be described as politics of despair. The PIITBY behavior emanates from a politics of compromise. The evasiveness of the NIMTOO and NIMEY behaviors also has negative connotations. These syndromes could be said to epitomize unresponsiveness on the part of politicians to the needs and concerns of citizens. Collectively, the politics from these syndrome behaviors can be viewed as a politics that results in winners and losers, or the politics associated with a zero sum game.

From the positive point of view, the politics characterized by WIMBY, NIMBY, and PIITBY behaviors demonstrate American politics at its best. The effectiveness of the NIMBY behavior demonstrates the power associated with the "politics of unity." NIMBY and WIMBY are also illustrative of what can be characterized as a "politics of community." The NIMTOO and NIMEY behaviors that constitute responses to this politics of community further evidence the power that resides within unity and community. The sense of accomplishment and possibility that emanates from these politics of unity and community often lead to a politics of empowerment, which in turn leads to a politics of hope. When individuals unite in response to an environmental threat, they form a community. The community, in turn, begins to feel empowered with the possibility of accomplishment. This opportunity for accomplishment can then lead them to hope for a brighter future, as illustrated by both Alsen and Clarion County.

SUMMARY

The response to potential environmental hazards is so intense that it has spawned a set of syndrome behaviors endemic to this body of politics. These syndrome behaviors represent action taken when citizens have been affected or think they are going to be affected by environmental decisions and actions by politicians who are forced to respond to citizens' concerns. Several of these behaviors have been discussed in this chapter. The impact that various syndrome behaviors have on the development and implementation of policy decisions surrounding environmental hazards has also been discussed. Five prevalent behaviors have been considered. Case studies illustrate behavioral characteristics associated with each syndrome. These syndrome behaviors and the politics of environmental justice that emanates from them are making important contributions to the American political system.

NOTES

1. David E. Camacho, *Environmental Injustices, Political Struggles: Race, Class, and the Environment* (Durham, N.C.: Duke University Press, 1998), 14.

2. Harold Dwight Lasswell, *Politics: Who Gets What, When, How* (New York: McGraw-Hill, 1936).

3. Lasswell, *Politics*.

4. Syndrome: a group of signs and symptoms that collectively indicate or characterize a disease, psychological disorder, or other abnormal condition.

5. Daniel Mazmanian and David Morell, "The 'NIMBY' Syndrome: Facility Siting and the Failure of Democratic Discourse," in *Environmental Policy in the 1990s,* ed. Norman J. Vig and Michael E. Kraft, 125–44 (Washington, D.C.: Congressional Quarterly Press, 1990).

6. Walter A. Rosenbaum, *Environmental Politics and Policy* (Washington, D.C.: Congressional Quarterly Press, 1991), 236.

7. Randall Ripley and G. A. Franklin, *Congress, the Bureaucracy and Public Policy* (Homewood, Ill.: Dorsey, 1980), 62.

8. Gerald Jacob, *Site Unseen: The Politics of Siting a Nuclear Waste Repository* (Pittsburgh: University of Pittsburgh Press, 1990), 149.

9. Jacob, *Site Unseen*.

10. Nicholas Lenssen, *Nuclear Waste: The Problem That Won't Go Away* (Washington, D.C.: Worldwatch Institute, 1991), 21.

11. Walter A. Rosenbaum, *Environmental Politics and Policy* (Washington, D.C.: Congressional Quarterly Press, 1991), 118.

12. J. M. Burn, J. W. Peltason, and T. E. Cronin, *Government by the People* (Englewood Cliffs, N.J.: Prentice Hall, 1985), 286.

13. Carol Barner-Barry and Robert Rosenwein, *Psychological Perspective on Politics* (Englewood Cliffs, N.J.: Prentice Hall, 1985).

14. Michael R. Edelstein, *Contaminated Communities: The Social and Psychological Impact of Residential Toxic Exposure* (Boulder, Colo.: Westview, 1988), 195.

15. Mary McCastle, interview by author, 2002.

16. "Rollins Targeted over Fumes That May Have Made Kids Ill," *New Orleans Times-Picayune,* February 12, 1986.

17. "Plant Fined by Judge over Odor," *New Orleans Times-Picayune,* October 1, 1985.

18. Steve Swartz, "Rollins Environmental in Fight over Plant: Regulators Seek to Close Toxic-Waste Facility," *Wall Street Journal,* March 28, 1986, 6.

7

Confronting Environmental Injustice in Connecticut

Mark Mitchell, Cynthia R. Jennings, and James Younger

The traditional approach to addressing environmental pollution under current federal and state legislation is to identify a potential pollutant, research it for years or decades until it is proven beyond reasonable doubt to be harmful, then identify sources of the pollutant and develop regulations to limit the pollution from each identified source. Although this approach has been effective in reducing the total emissions of these specific pollutants, it has several limitations. Since it requires that substances be proven unsafe before they can be regulated, it is a conservative approach to protecting human health and the environment. The process is slow and deliberate and may not reflect current scientific information. New chemicals are being produced faster that they can be regulated and the process does not account for cumulative risk of exposure to multiple toxic substances. There are no provisions to limit toxic exposure to the fewest number of people or to account for differing susceptibilities of populations to health effects from a particular exposure. There is little chance to adapt regulation of a pollution source to the peculiarities of a specific situation and the provisions do not adequately prevent vulnerable populations from being disproportionably burdened with negative environmental impacts, such as increased exposure to pollution and less access to environmental amenities.

For these reasons, government, academics, scientists, and community organizations are recognizing that community-based environmental protection can play a key role in helping society achieve environmental justice. The sections that follow detail three different approaches taken to confront environmental injustice in Connecticut, from adopting an innovative policy at the federal level, to advocating for environmental justice through community organization, to being that single voice of reason that points out the dangers of an environmentally risky business.

* Section 1. In "EPA New England's Urban Environmental Initiative: A Case Study of Community-Based Environmental Protection," James Younger, director of the Office of Civil Rights and Urban Affairs at the U. S. Environmental Protection Agency (EPA) New England, describes how EPA is moving to incorporate community-based environmental protection programs as additional tools to address environmental injustice. Specifically, Younger details an innovative regional EPA pilot program to tackle urban environmental and associated public health problems.
* Section 2. In "Multiracial, Cross-Cultural Environmental Mobilization: One Person Can Make a Difference," Cynthia Jennings, board chair of the Connecticut Coalition for Environmental Justice (CCEJ), describes how she started the environmental justice movement in Connecticut by organizing her neighbors to fight against a proposed landfill expansion.
* Section 3. In "Protecting Urban Environments in Connecticut through Community Education and Advocacy," Dr. Mark Mitchell, president of the Connecticut Coalition for Environmental Justice, describes how the three principles of the organization—research, education, and advocacy—have improved the environment, health, and quality of life of urban residents. Mitchell reveals some of Connecticut's discriminatory siting policies for environmentally risky facilities and shows how community organizations can effectively combat such practices.

EPA NEW ENGLAND'S URBAN ENVIRONMENTAL INITIATIVE: A CASE STUDY OF COMMUNITY-BASED ENVIRONMENTAL PROTECTION

The U.S. Environmental Protection Agency (EPA) has traditionally used regulatory and enforcement policy to address environmental problems, often described as a "command and control" approach. Increasingly, however, the EPA is recognizing that the regulatory and enforcement framework does not adequately address some existing and emerging environmental problems, especially those that are site-specific or have more than one cause. The EPA has begun approaching environmental concerns that do not fit the traditional framework using what the agency terms "community-based environmental protection." Using this approach, the EPA brings together public and private stakeholders to assess community needs and then works to provide resources that enable the community to build local capacity to identify and resolve issues that may have a negative impact on the environment, public health, and overall quality of life. As a leader in implementing EPA's environmental justice policy, the New England office has successfully used community-based environmental protection as a guiding principle in its efforts to address environmental justice concerns.

In 1993, EPA New England issued an environmental equity policy to address what is now commonly referred to as environmental justice, becoming the first EPA regional office to do so. Since then, EPA New England has made fundamental changes in the way everyday work is carried out by realigning staff and resources to better enable the agency to work with the public. The region's most recent Environmental Justice Policy and Action Plan, issued in August 2001, goes even further by charging the office directors with the responsibility for developing environmental justice policy, guidance, and implementation strategies to institutionalize environmental justice activities throughout the EPA New England office.

EPA defines environmental justice as the fair treatment and meaningful involvement of all people, regardless of race, color, national origin, or income, with respect to the development, implementation, and enforcement of environmental laws, regulations, and policies. Fair treatment means that no group of people, including any racial, ethnic, or socioeconomic group, should bear a disproportionate share of negative environmental consequences resulting from industrial, municipal, and commercial operations or the execution of federal, state, local, and tribal programs and policies. EPA New England strives to involve minority and low-income communities in its decision-making processes because, while these communities often suffer a disproportionate negative impact from environmental pollution and threats to public health, they may not have access to resources to get their concerns addressed.

One program that exemplifies how EPA New England uses community-based environmental protection to further its environmental justice goals is the Urban Environmental Initiative (UEI) pilot program. The UEI was launched in 1995 to address disproportionate environmental and public health problems in the targeted cities of Boston, Massachusetts; Providence, Rhode Island; and Hartford, Connecticut. The program facilitates community-based environmental protection in these cities by taking an active role in listening to community needs and concerns, identifying projects for joint action, and providing resources to implement projects that make measurable improvements in public health and the quality of the urban environment. To support the UEI program, EPA New England dedicates one full-time manager to each of the target cities and provides financial resources including grant money to fund program activities.

In Hartford, the city's UEI program manager began assessing community needs by attending public forums and forming a critical partnership with a local community group, ONE/CHANE, Inc. (Organized Northeasterners/Clay

Hill and North End), to prevent the expansion of a local landfill. Since 1995, the Hartford UEI program has grown considerably and now works with almost forty public, private, and nonprofit groups to address community environmental and public health problems. Some of the Hartford UEI program's most recent accomplishments include reclaiming and converting a two-acre contaminated vacant lot into a clean, passive park and community garden; convening an Asthma Policy Forum and Legislative Briefing to engage and inform the public and provide guidance to local policy makers to address Hartford's asthma emergency; and supporting efforts to convert an overgrown area along the South Branch of the Park River into a productive urban greenway.

EPA New England's UEI pilot programs in Hartford and other cities demonstrate that together public, private, and nonprofit groups can achieve measurable environmental results and help reverse the disproportionate risks that some low-income and minority communities have faced for decades. The challenge facing stakeholders now is to learn how to build on these small successes to secure equity and environmental justice for all in the future.

MULTIRACIAL, CROSS-CULTURAL ENVIRONMENTAL MOBILIZATION: ONE PERSON CAN MAKE A DIFFERENCE

For years, residents on the north side of Hartford, Connecticut, were assaulted by putrid, acrid odors from the local landfill. This landfill was owned by the city of Hartford, but operated by a quasi-governmental agency, the Connecticut Resources Recovery Authority, which brought waste from more than sixty towns in four states to the North Hartford Landfill. Government officials and the landfill operator had been unresponsive to complaints about the odors from the mostly low-income, black, and Latino residents of this neighborhood.

As a lifetime resident of North Hartford and a block captain for our neighborhood organization, ONE/CHANE, Inc., I was appalled to learn that the state was considering expanding this landfill, which had created so many problems in our community. No one in the North Hartford community knew of the proposed landfill expansion in the community until we read about it in the local newspaper the day before the State Department of Environmental Protection (DEP) hearings were scheduled to take place. I contacted the executive director of ONE/CHANE, Inc., and several other neighbors and elected officials who, on one day's notice, filled a bus to go attend the public hearing about the landfill expansion. None of the thirty-three people present had ever been to a public hearing before, but we were not intimidated. It was clear from the response of DEP officials that they had never seen thirty-three angry black people at an environmental hearing. We all signed up to speak and we all opposed the expansion of the landfill on the record, because we believed that the landfill was detrimental to our community and would be even more detrimental if expanded.

We did not trust the landfill operators, we did not trust DEP officials, and we had no idea what procedures we would have to go through to stop the landfill expansion. No one provided information to us prior to the public hearing, and the information that they gave us during the hearing was of no help because we did not understand the public hearing process. We were later told that if we wanted to intervene we would need to retain an attorney. We could not afford an attorney and we did not know any environmental attorneys. We didn't even know what it meant to intervene. One member of our block association called someone they knew who was an environmental activist at the other end of the state and he was able to secure an environmental attorney to assist us. In the meantime, we contacted our town committee, our city council members, our state legislators, our congressional representatives, the DEP, our state health department, and the director of the Hartford Health Department.

Our block association began meeting with other residents and community organizations to obtain assistance in opposition to the landfill expansion. One city council member drafted a resolution to oppose the expansion, but city attorneys advised the council members that they were prohibited from passing such a resolution by contractual agreement with the landfill operator. This left us on our own to oppose the expansion, so ONE/CHANE, Inc., sponsored a community forum on the North Hartford Landfill. More than 250 people, including the attorney general of the state of Connecticut, U.S. Justice Department officials, congressional staff, state legislators, city council members, and staff from the EPA, DEP, and state and local health departments attended. Out of this community forum came a commitment from the congressional representatives to provide us with assistance in obtaining a health assessment from the U.S. Agency for Toxic Substances and Disease Registry. Each of the agencies in attendance at the community forum assisted in educating the community about environmental justice, environmental health, and other issues surrounding the landfill.

The DEP issued a consent decree with more than thirty operational violations that the landfill operator was required to correct before they would consider granting an expansion permit. This consent decree resulted in the landfill operator cleaning up the landfill, reducing odors, and installing safety devices as requested by the community. Through ongoing community education and the support of EPA New England staff, we created a general base of knowledge regarding the environment, health, and environmental justice in Hartford. Because of this heightened awareness, the community was ready to respond when the next environmental justice insult occurred. In 1997, the local utility company attempted to permanently site a jet-engine power generator in Hartford, as a partial replacement to the closing of all four nuclear power plants in rural Connecticut. This triggered the formation of the Hartford Environmental Justice Network, a coalition of neighborhood organizations and community groups from throughout the Hartford area. This group was a cross-cultural, multiracial, and geographically diverse coalition that built on the knowledge and awareness generated by ONE/CHANE, Inc., to successfully fight the siting of this new power plant. This network has expanded to more than thirty organizations and continues to address a number of environmental justice issues facing Hartford. These issues include the banning of medical waste disposal in Hartford, accountability for a sewage sludge compost fire that burned for more than eight days, coordinating a public hearing about dioxin, declaring an asthma emergency in Hartford, and stopping the siting of Connecticut's largest diesel truck stop in Hartford.

Every environmental justice issue begins with one primary actor, one person who may believe that something is inherently wrong with a facility, with public health, or with the air or water. Many environmental justice matters are identified purely by accident, because one individual becomes convinced that something is wrong or that an environmental injustice is occurring. When you have issues related to environmental justice, and when people of color are predominantly being adversely impacted by an environmentally risky facility, everyone in the state is affected. People of color are not the only people living in the urban centers. Many white residents continue to reside in the cities and are as severely affected as blacks, Latinos, and other people of color. In order to bring about an effective environmental justice movement, we must reach out to all affected constituencies and involve them in the environmental justice movement. Polluted air does not stop at the city line. If we are going to meet global environmental goals and objectives, we must reduce pollution in the urban centers.

PROTECTING URBAN ENVIRONMENTS IN CONNECTICUT THROUGH COMMUNITY EDUCATION AND ADVOCACY

In Connecticut there is a direct correlation between the percentage of people of color in a town and the number of potential environmental hazards in that town (fig. 7.1).[1] In fact, the correlation between the percentage of people of color and the number of potential environmental hazards is stronger than the relationship between the percentage of low-income residents and the number of potential environmental hazards in a given community. This correlation is defined as environmental racism.

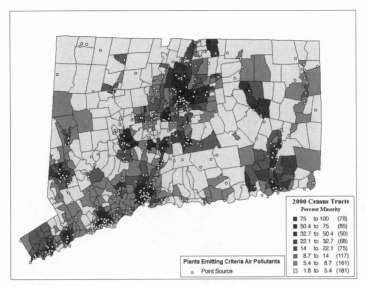

Figure 7.1.   2000 Census Tracts: Percent Minority and Plants Emitting Criteria Air Pollutants

Hartford produces more than three thousand tons of point source air pollution annually (fig. 7.2). More than half of it comes from the burning of trash from sixty-six Connecticut towns in the Hartford trash-to-energy incinerator. The largest sewage sludge incinerator in the state and two smaller power plants located within the city limits produce another 40 percent of Hartford's point source air pollution. In total, more than 90 percent of Hartford's point source air pollution comes from businesses that support regional trash and energy needs. These facilities unnecessarily expose large numbers of vulnerable Hartford residents to health risks from air pollution and air toxins. Similarly, more than 90 percent of New Haven's and Bridgeport's point source air pollution comes from regional power generators and regional trash incineration (figs. 7.3 and 7.4). By contrast, Waterbury, the fourth-largest city in Connecticut, produces only 116 tons of point source air pollution per year (fig. 7.5). Most of this air pollution is from businesses that support the local economy.

In 1997, a third power generator was sited adjacent to a Hartford neighborhood where 80 percent of the residents are black or Latino and that has high rates of asthma. The new jet-engine power generator was built in the city without community notification or involvement, and neighborhood residents questioned why another facility that is known to trigger attacks in people with asthma was sited within two miles of an estimated 10,000 people known to suffer from asthma. The Hartford Environmental Justice Network (HEJN) was formed to address what was perceived as discriminatory siting of an environmentally risky facility, without appropriate public notification or involvement.

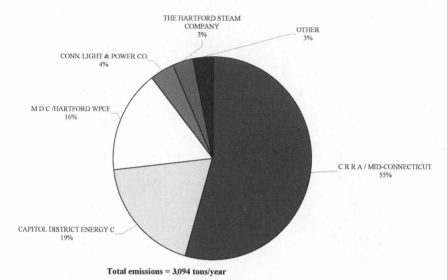

THE HARTFORD STEAM
COMPANY
3%

OTHER
3%

CONN. LIGHT & POWER CO.
4%

M D C /HARTFORD WPCF
16%

C R R A / MID-CONNECTICUT
55%

CAPITOL DISTRICT ENERGY C
19%

Total emissions = 3,094 tons/year

Figure 7.2.   Hartford Non-Traffic Air Pollution Sources 1999
Source: Connecticut Department of Environmental Protection

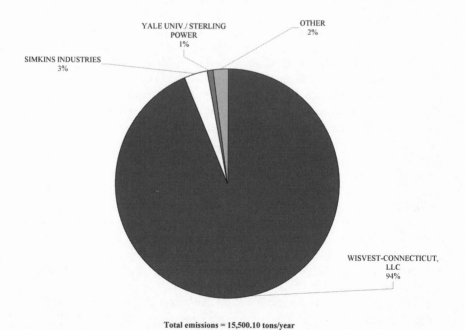

YALE UNIV./ STERLING
POWER
1%

OTHER
2%

SIMKINS INDUSTRIES
3%

WISVEST-CONNECTICUT,
LLC
94%

Total emissions = 15,500.10 tons/year

Figure 7.3.   New Haven Non-Traffic Air Pollution Sources 1999
Source: Connecticut Department of Environmental Protection

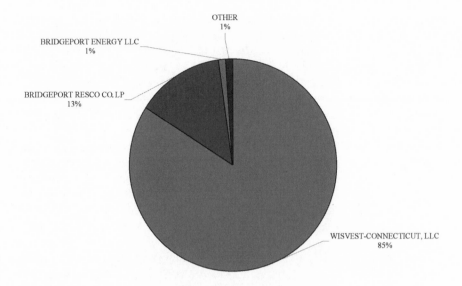

Total emissions = 14,913 tons/year

Figure 7.4.   Bridgeport Non-Traffic Air Pollution Sources 1999
Source: Connecticut Department of Environmental Protection

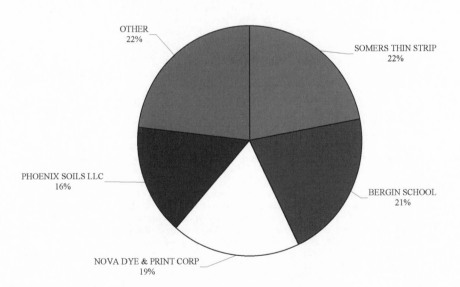

Total emissions = 115.61 tons/year

Figure 7.5.   Waterbury Non-Traffic Air Pollution Sources 1999
Source: Connecticut Department of Environmental Protection

The HEJN was able to organize eight neighborhood, business, and community organizations that persuaded the utility company to shut down the new power plant. This may have been the first time a community was able to negotiate the closing of a newly operational power plant.

HEJN's success in negotiating the closing of the power plant marked the beginning of the organization's effort to block disproportionate siting of potential environmental hazards in Hartford. The following year, HEJN expanded statewide to form the Connecticut Coalition for Environmental Justice (CCEJ). Since 1998, the CCEJ has grown to encompass more than fifty organizations and has established local chapters in Hartford and New Haven. The CCEJ's mission is to protect urban environments in Connecticut through community education, advocating changes in state policy, and promoting individual, corporate, and governmental responsibility for the environment (fig. 7.6). The CCEJ's definition of environment encompasses places where we live, work, play, and go to school.

Specifically, the CCEJ helps low-income communities and communities of color advocate for the elimination of discriminatory siting policies for

---

Connecticut Coalition for Environmental Justice

Mission Statement

The mission of Connecticut Coalition for Environment Justice is to protect urban environment in Connecticut through educating our community, through promoting changes in state policy, and through promoting individual, corporate and governmental responsibility towards our environment. We define environment as including the places that we live, work, play, and go to school. Specifically, we are educating and promoting policies to:

1. Eliminate discriminatory siting policies for production or storage of environmental toxins.
2. **Reduce** rates of occurrence of **environmental associated diseases** such as asthma and lead poisoning in the urban areas and communities of color in Connecticut.
3. **Ensure community notification** and involvement in the decision making process regarding issues and events that may significantly impact the urban environment.
4. **Reduce the burden of environmental toxins** in the urban areas and communities of color in Connecticut.
5. **Promote community involvement in decision making** for urban land use and to balance the development of physical, social as well as economic well-being of urban communities of color.
6. **Promote greater community benefits** for urban areas that bear a disproportionate burden of facilities and situations that reduce the quality of life in these communities. Such benefits may include community monitoring, employment opportunities, and funding.
7. **Educate people about the relationship between urban ecology and health, nutrition, economics, and well being.**
8. **Educate the public about the disparities in environmental burdens** borne by urban communities and communities of color in Connecticut and the reasons why negative health impacts are greater in these communities.
9. **Create access to safe, healthy, clean, and accessible environmental amenities** and recreation for the people of our communities in parks, rivers, and outdoor activities.
10. **Create opportunities** for young people to participate in environmental educational and recreational activities.
11. **Conduct and coordinate research** and studies related to the urban environment and environmental health and safety.
12. **Educate our communities to recognize and respect our traditional relationship with Mother Earth** and to promote sustainable development of Her resources to that our children can continue to enjoy the benefits and protection that She has provided to us.

Figure 7.6.   Connecticut Coalition for Environmental Justice Mission Statement

environmentally hazardous facilities, for example, policies that promote the location of new regional waste facilities adjacent to other regional waste facilities. In addition, the CCEJ opposes an energy deregulation law that was passed in Connecticut in 1998. This law exempts any power plant from the requirement to conduct an environmental impact assessment if it is sited on land where a power plant had been located at any time in the past. Therefore, if one hundred years ago there was a small electric generator on the property next door to your house, a new large power plant can locate there now without assessing the environmental impact of such a facility.

The CCEJ strives to ensure that the community is notified and involved in decision-making processes that may significantly impact the urban environment. Community notification processes that work well in suburban and rural areas are often not effective in notifying urban residents. For example, placing a legal notice in a two-hundred-page daily urban newspaper may not notify as many people as the same legal notice published in a twenty-page weekly suburban newspaper. Community notification is further limited because many urban residents are not native English speakers and there are no requirements to notify residents in their primary language.

Besides environmental hazards associated with discriminatory siting, urban residents often receive no benefit from the facilities that are placed in their neighborhoods. A new company that moves into the neighborhood may not hire neighborhood residents. The CCEJ educates community residents to ask for benefits that will improve the quality of life and economy of their neighborhoods and minimize any negative consequences of potential environmentally hazardous businesses. Community education is the key to empowering urban residents to advocate for the elimination of discriminatory siting and to tackle general community environmental and public health problems. The CCEJ educates residents about the scientific research on the relationship between environmental hazards and associated diseases such as asthma, cancer, lead poisoning, and even diabetes. Both environmental hazards and incidences of these diseases occur disproportionately in communities of color, and the CCEJ believes this is not a coincidence.

The CCEJ teaches residents of low-income communities and communities of color about the relationship between health, nutrition, economics, ecology, and well-being. One example of this relationship is that children who are iron- or calcium-deficient will absorb more lead from paint than well-nourished children exposed to the same amount of leaded paint. Childhood lead poisoning is associated with a higher incidence of learning and behavior problems and may increase the high-school dropout risk. The CCEJ also educates youth in low-income communities and communities of color about the connection between the environment and their lives. The CCEJ teaches residents the importance of recycling by explaining that the trash they throw away will probably be burned in Hartford and may affect the air that they breathe. Another focus for CCEJ is creating recreational opportunities for urban youth so they can appreciate natural resources.

The CCEJ promotes research that will address environmental health issues of concern to urban residents, focusing on certain diseases and health conditions that are much more common in urban areas than elsewhere. Mainstream researchers often overlook these diseases and conditions, yet these conditions are often the beginnings of national trends. For example, Hartford residents and urban residents throughout the country complain of an upper respiratory condition resembling a cold that lasts longer than one month. Research shows that 21 percent of Hartford residents complain of a cough and sore throat lasting longer than two weeks. This condition, which the CCEJ terms "Chronic Recurrent Respiratory Ailment," seems to be distributed in the same pattern as asthma in Hartford and is not described in medical literature. The CCEJ also educates urban residents to recognize and respect the traditional relationship between people of color and Mother Earth and to promote sustainable development of her resources. The CCEJ believes that chemicals should be proven safe before they can be released into the environment. Currently, few new chemicals are required to be tested for health effects prior to release.

Following these principles of research, education, and advocacy, the CCEJ has been instrumental in accomplishing many environmental goals throughout Connecticut, including blocking asphalt plants from being located in residential neighborhoods in Bridgeport, banning medical waste disposal facilities in Hartford, requiring the sewage sludge incinerator in New Haven to monitor dioxin emissions annually, stopping the location of the largest truck stop in New England in Hartford, requiring the oldest and dirtiest power plants in Connecticut to substantially reduce air pollution, and, in response to a survey of eight thousand children showing that 41 percent of all children and 48 percent of Latino children studied have asthma, persuading the city of Hartford to declare an asthma emergency.

When HEJN was formed, Hartford was importing trash from Vermont, Massachusetts, and New York City. There was more New York City trash being burned in Hartford than Hartford trash. The Hartford trash-to-energy facility is no longer accepting out-of-state trash and has reduced its emissions of nitrogen oxides and dioxin. These improvements were made possible by educating and empowering community residents to take action. The CCEJ accomplishments are a testament to what a powerful tool research-based community education and advocacy is for improving the urban environment, but the organization believes there is much more work to be done.

CONCLUSION

There are many approaches to initiating change to protect low-income communities and communities of color from environmental injustice, ranging from reactions by an individual citizen to the dangers of a local environmental hazard to the enactment of policies at the federal, state, or local level that broadly protect groups of citizens from disproportionate negative environmental risks. We can also see that community-based environmental advocacy can be an effective tool in achieving environmental justice. Although this advocacy is necessarily born of passion, it requires resources and information to sustain itself and flourish. Government, academia, scientists, and communities all have key roles in supporting community-based environmental advocacy to protect our most vulnerable population from environmental injustice.

RECOMMENDED READINGS

Bullard, Robert D. *Dumping in Dixie: Race, Class, and Environmental Quality.* Boulder, Colo.: Westview, 1990.

Edwards, Audrey. "Programs That Work." *Essence* 28, no. 3 (July 1997): 42.

Ember, Lois. "IOM Report Plumbs Environmental Justice Issues." *Chemical & Engineering News* 77, no. 11 (March 1999): 14.

Fritz, Jan Marie. "Searching for Environmental Justice: National Stories, Global Possibilities." *Social Justice* 26, no. 3 (Fall 1999): 174–89.

Fugazzotto, Peter. "Angling for Environmental Justice." *Earth Island Journal* 9, no. 3 (Summer 1994): 19.

Hines, Revathi I. "African Americans' Struggle for Environmental Justice and the Case of the Shintech Plant: Lessons Learned from a War Waged." *Journal of Black Studies* 31, no. 6 (July 2001): 777–89.

Marquez, Benjamin. "Mobilizing for Environmental and Economic Justice: The Mexican-American Environmental Justice Movement." *Capitalism, Nature, Socialism* 9, no. 4 (December 1998): 43–60.

Moberg, Mark. "Co-Opting Justice: Transformation of a Multiracial Environmental Coalition in Southern Alabama." *Human Organization* 60, no. 2 (Summer 2001): 166–77.

Novotny, Patrick. "Where We Live, Work and Play: Reframing the Cultural Landscape of Environmentalism in the Environmental Justice Movement." *New Political Science* 32 (Summer 1995): 61–79.

Schlosberg, David. "Networks and Mobile Arrangements: Organizational Innovation in the US Environmental Justice Movement." *Environmental Politics* 8, no. 1 (Spring 1999): 122–48.

Smith, Rhonda. "MOSES Leads Winona, Texas, to Environmental Justice." *Crisis (The New)* 104, no. 1 (July 1997): 30–31.

Terry, Larry. "Activism on the Bayou." *Chemical Week* 161, no. 20 (May 1999): 59.

NOTE

1. See the report by Ross Bunnell, "A Preliminary Analysis of the Distribution of Various Environmental Sites in Connecticut with Respect to Race and Ethnicity" (Hartford: Connecticut Department of Environmental Protection, April 1997), and Timothy Black and John A. Stewart, "Burning and Burying in Connecticut: Are Regional Solutions to Solid Waste Disposal Equitable?" *New England Journal of Public Policy* (Spring/Summer 2001): 15–34.

8

For the People: American Indian and Hispanic Women in New Mexico's Environmental Justice Movement

Diane-Michele Prindeville

THE ENVIRONMENTAL JUSTICE MOVEMENT IN NEW MEXICO

The environmental justice movement is active in New Mexico in local, state, and tribal politics. The state's history, political culture, large racial/ethnic minority population, political economy, and location have all contributed to the founding and growth of environmental justice groups throughout New Mexico. The legacy of colonization, first by Spain and then by the United States, has left an indelible mark on New Mexico's culture, economy, and politics. Unresolved conflicts involving Indian sovereignty, Spanish land grants, the U.S. government, and private developers continue to fuel debates over rights to land and water.[1] Tensions persist between the economic draw of tourism on the one hand and the commercialization of indigenous peoples and their cultures on the other. What's more, New Mexico's economic dependence on both the federal government and the defense industry has left a legacy of environmental destruction. Specifically, hazardous and nuclear wastes, produced through the mining of uranium and the manufacture and testing of nuclear weapons in the state, have led to countless deaths due to radiation exposure and have resulted in significant environmental contamination of soil and groundwater.[2] In addition, the Waste Isolation Pilot Project (WIPP), a national storage site for medium- and low-level radioactive wastes, has been constructed in New Mexico, despite considerable protest. The combination of these factors and their social, economic, and political impact serve to make the environmental justice movement's agenda especially relevant to New Mexicans. Consequently, grassroots organizations have formed throughout the state in both urban centers and rural areas, on and off Indian lands.

These grassroots groups have grown, building coalitions with other organizations and forming what has become known as the environmental
justice movement. In New Mexico, American Indian and Hispanic women
have been particularly active in this movement and make up a significant
portion of the leadership. In the early 1990s, for example, members of the
Navajo Nation and the Mescalero Apache tribe successfully defeated proposals to place hazardous waste storage facilities on their tribal lands. In
both cases, the grassroots campaigns were spearheaded by women. Also
during this period, under the leadership of its first female governor, Isleta
Pueblo adopted its own nationally recognized environmental standards in
order to combat pollution generated by the city of Albuquerque. Similarly,
neighborhood groups in Albuquerque's South Valley, a predominantly Hispanic area, pressured the city to investigate the source of groundwater contamination that resulted in a number of wells being capped. When the pollution was found to emanate from the nearby weapons research laboratory
located at Kirtland Air Force Base, community leaders, with the help of a
local environmental justice organization, managed to negotiate the cleanup
of several toxic sites.

    Like similar groups around the country, New Mexico's environmental
justice organizations identify concerns arising from environmental and
economic conditions, such as community development, neighborhood
safety, pollution, the availability of low-income housing and public transportation, employment opportunities, and workplace hazards.[3] Local
groups have campaigned to influence public policies involving the siting
of undesirable land uses,[4] groundwater contamination,[5] industrial air
emissions and effluent discharges,[6] airport noise, soil erosion and toxicity,[7] and waste reduction and incineration.[8] Additionally, these organizations are involved in issues surrounding neighborhood and cultural
preservation,[9] Indian sovereignty,[10] indigenous people's rights,[11] and civil
rights abuses. These grassroots organizations form coalitions with other
groups in a variety of activities to protect sacred Indian lands,[12] to protest
anti-immigration or English-only laws, to boycott union-busting companies, or to advocate for community health and safety.[13] Besides respond-
ing to problems, these organizations work proactively to highlight citizen
concerns and to develop solutions.[14]

    In general, New Mexico's grassroots environmental groups (1) endorse
policies that favor the disenfranchised, (2) focus on equality and distributional impacts, (3) advocate direct action, and (4) solicit the support of local
civic and religious groups.[15] While they vary with regard to membership, objectives, and strategies, environmental justice groups in New Mexico share
broad goals and policy agendas with similar organizations around the country.[16] In fact, numerous New Mexico groups maintain ties with organizations
that also seek environmental, economic, and social justice for indigenous

peoples across the United States and in other countries. For instance, the Southwest Network for Environmental and Economic Justice (SNEEJ) was established in 1990 as a regional multi-issue coalition by the SouthWest Organizing Project, a vocal social justice group based in Albuquerque. The following "Statement of Solidarity" was presented by SNEEJ leaders at the People of Color Regional Activist Dialogue held in Albuquerque the same year. It outlines both the group's mission and the precepts underlying the environmental justice movement.

> We are a multi-cultural, multi-national, grassroots network whose focus is to address the fact that communities of color, as well as economically oppressed communities, suffer disproportionately from toxic contamination. We are deliberately targeted through genocide of indigenous people, the threatening of future generations, racism, sexism, and a lack of economic and social justice. (SNEEJ, April 8, 1990)

For the most part, New Mexico's numerous grassroots environmental organizations are neither as radical as Earth First! nor as tied to the establishment as the Sierra Club.[17] Instead, they tend to reflect local racial and class diversity, offering opportunities for political involvement to people of color and the poor, which these other associations do not.[18] Individuals become involved and remain active because of the relevance of the problems being addressed: jobs, health, family, and neighborhood safety.

WOMEN'S POLITICAL ACTIVISM AND THE ENVIRONMENTAL JUSTICE MOVEMENT

Historically, social movements have been a primary vehicle for women's political participation.[19] As one scholar notes, "In the procedural or pluralistic democracy of the United States, organizations and organized interest groups represent the major vehicle for input into decision making regarding policy and resource allocations at all levels of government."[20] Successful citizen organizations enable individuals to effect change through collective action. In doing so, they cause a shift from the impersonal, unilateral decision making of politicians, businesses, and technicians to mediated resolutions arrived at through negotiation with community leaders. This form of grassroots involvement is particularly accessible to women who are already a part of community networks that provide resources, support, and communication links with other residents. These networks are familiar and informal. They build on existing relationships established among women through their neighborhoods, children's schools, places of worship, clubs and civic organizations, and places of work.[21] Furthermore, community organizing activities employ many of the skills that women develop through their roles as homemakers and mothers. Women's involvement in community-based groups also provides occasions for social interaction, status and recognition, an increased sense of personal and political efficacy, confidence building, and greater personal satisfaction.[22]

Through their participation in local grassroots associations, women create a legitimate voice to represent their interests in the governing bureaucracy, what one scholar calls "activist mothering."[23] When such organizations are managed and led by women, they often cultivate local women's leadership, provide greater opportunities for women's involvement, and enable participants to develop skills transferable to other forums. As an avenue for political participation, grassroots organizing has proven to be especially accessible to women, often leading to greater opportunities for leadership.[24] While women are not proportionally represented in highly visible leadership positions within the environmental justice movement, it is estimated that 90 percent of those who actively participate are women.[25] In fact, a great many environmental justice groups in the United States and abroad are founded and run by women of color.[26] These activists, who tend to work through structurally and ideologically democratic organizations to improve conditions in their communities,[27] are often mothers fighting problems that threaten the health and well-being of their families and neighborhoods.[28] Their motivation, however, is not solely a product of their biological or gendered sociocultural roles of "mother" and "nurturer." Rather, their motives for involvement in such grassroots politics, and in electoral politics, are far more complex. As G. Di Chiro notes,

The question of community survival in the face of cultural imperialist attacks by the dominant, white male, industrial complex figures conspicuously in many women of color's involvements in environmental justice works. Women in many communities and cultures have customarily been seen to be the repositories of or given the responsibilities for maintaining local, cultural traditions and histories.[29]

Both American Indian and Hispanic women activists have traditionally engaged in struggles to remedy social and economic concerns; this work continues through their leadership in contemporary environmental justice organizations. Whether they are native women fighting the federal government over deployment of nuclear weapons on tribal lands or Mexican American mothers organizing against the placement of yet more undesirable land uses in their *barrio,* they share common approaches to leadership and their activism is often similarly motivated. The focus of this chapter is the role of American Indian and Hispanic women in New Mexico's environmental justice movement and their contributions to the political, social, and economic well-being of their communities.

RESEARCH METHODOLOGY

Sample Selection

Using nonrandom purposive sampling, a total of fifty women active in New Mexico politics were selected for participation in this study. I obtained their names using a reputational "snowball" technique in which each woman interviewed was asked the names of other women involved in environmental politics in the state. The bulk of the participants (forty-five) were leaders known for their involvement in environmental policymaking or for their roles as advocates for communities of color. The remaining five leaders favored environmental and social justice policies, but their primary political involvement was with other issues. The reliability and validity of my sampling strategy was reinforced when the same women were repeatedly identified as policy leaders by different study participants in numerous organizational settings. Generalization to the larger population of women leaders across the United States is not intended, nor would it be appropriate, using this methodology.

Both indigenous and Hispanic women were interviewed for this project, including volunteers and paid staff of grassroots organizations (activists) as well as appointed and elected officials at various levels of government (public officials).

Table 8.1 provides a breakdown of the racial/ethnic identity of the leaders interviewed as well as the positions they held. While the indigenous leaders were evenly split between activists and officials, among Hispanics there were two more grassroots activists than public officials. Overall, though, the study participants were fairly evenly distributed among staff (thirteen) and volunteers (thirteen) of grassroots organizations and appointed (twelve) and elected (twelve) public officials.

American Indian women and women of Hispanic origin eighteen years and older make up roughly 313,500 persons or about 17 percent of New Mexico's population.[30] While the precise number of people who hold leadership positions in communities throughout the state is unknown, we can safely assume that women are a minority within this elite group. Subsequently, the small

Table 8.1. Racial/Ethnic Identity of Leaders and Position Held

Race/Ethnicity	Grassroots Activists		Public Officials		Total
	Staff	Volunteer	Appointed	Elected	
Indigenous[1]	6	7	7	6	26
Hispanic[2]	7	6	5	6	24
Total	13	13	12	12	50
Percent	26	26	24	24	100

[1] This group consists of twenty-five American Indian women and one Native Hawaiian.
[2] This group of Hispanas includes two women who self-identify as "Mestizas," acknowledging both their Native American and Spanish heritage.

universe from which to draw a sample resulted in my interviewing many, if
not most, of the indigenous and Hispanic women leaders in New Mexico politics. Due to their relatively small number, high level of political activity, and
aggressive coalition building, and because they constitute a political elite, the
women interviewed for this study were in many cases acquainted with each
other. To protect their identity, and so that no quote is directly attributable to
any individual, pseudonyms have been used throughout.

The twenty-six grassroots activists included employees and volunteers
who were, or had been, policy makers, such as directors or board members
in nonprofit, community-based organizations working for environmental
and social justice. Their organizations varied in scope as well as focus. Those
with a local scope served a particular neighborhood or an entire city. Some
groups worked throughout the state of New Mexico, while others were limited to serving the members of particular Indian tribes. Some of the organizations had a larger geographic scope that included several states within the
Southwestern region of the country. A few of the organizations represented
groups throughout the nation, while others reached across national boundaries to serve even broader constituencies.

Several of the public officials had gained valuable political experience as
former community activists. The public officials included elected and appointed officials who had campaigned on issues related to environmental
quality or who had served on formal legislative committees or in federal,
state, tribal, or local agencies addressing environmental and social justice
concerns. The public officials were a very diverse group comprising elected
or appointed leaders at the local, county, state, tribal, and national levels of
government. While most of the officials held positions in the legislative or
executive branches, two women worked in the judicial branch. All twenty-
four public officials personally advocated environmental or social justice
policies.

Data Collection and Analysis

Using an interview guide composed of open-ended questions to collect
my data, I met with the leaders at various locations throughout New Mexico
between April 1994 and June 1996 (with the exception of three interviews
conducted in November 1991 as part of a pilot project). Interviews were tape
recorded with the permission of the participant and handwritten notes were
taken simultaneously. A transcribed record of each interview was later made
from my field notes and audiotapes. The interviews lasted between approximately forty-five minutes and two hours, with an average length of one
hour.

Two interview guides were prepared, one for the activists and one for the
public officials, with only minimal variations to account for the leaders' dif-

ferent circumstances. While the interview guides followed a similar set of questions, participants spoke in their own words and were encouraged to pursue related issues expounding on areas they thought relevant. Both the format of the guide and the interview process stressed participants' definitions of concepts and issues, encouraged them to structure their own accounts of a problem or policy concern, allowed for detailed descriptions by the participants, and facilitated their communicating their particular ideas of what is relevant.[31] The flexible interview format gave participants special, nonstandardized treatment consistent with the research goal of developing as rich and as accurate a profile as possible of each of the leaders.[32] The use of loosely structured interviews, administered in person and employing open-ended questions, emphasizes the life experience of project participants. This technique allows respondents to tell their own stories, to offer their own subjective meanings, and to link these meanings to their understandings of the social world.[33] The flexibility afforded by this method allows for continual refinement of the interview schedule, clarification of questions and responses (because the researcher continually moves back and forth between data collection and analysis), inclusion of field observations in the analysis, and the exploration of previously unexamined concepts.[34] The result is rich data that reveal the complexities of human experience and emotion providing context, authenticity, specificity, and vivid description.

DISCUSSION OF THE FINDINGS

By obtaining mass support and adopting active leadership roles, indigenous and Hispanic women in New Mexico's environmental justice movement influence public policy and shape the politics of the state. The political activism of the indigenous and Hispanic women leaders I interviewed spans grassroots and electoral politics. While the New Mexico leaders' motives may incorporate traditional maternal and domestic concerns, the reasons for their political involvement are not limited to their gendered identities as women, wives, or mothers. Instead, their involvement in politics originates from many sources, including their early political socialization, their particular experiences of politicization, their affinity with and commitment to their racial/ethnic communities, their identification with and desire to improve the condition of women as a group, and their political ideologies.

Paths to Leadership

Although there is some overlap, the paths to leadership taken by the fifty women interviewed can be traced to three general sources. First, women mobilized as youths participated in social justice causes while in high school or college. Their exposure to varied political ideologies and their involvement in social movements and party organizations provided them with contacts and valuable experience. In many cases, this experience helped the young women to determine their future career goals and life choices, cementing their commitment to political activism. Second, while fulfilling their roles as mothers and homemakers, several women developed valuable planning and organizational skills, which aided their ascent to leadership. Women active in their children's education found occasions to influence school policy and programs. In numerous cases, their experiences led to their serving on local or state school boards and these positions led to yet other opportunities for political involvement. Third, women free to volunteer their time initiated or joined efforts to protect, maintain, or enhance the quality of life in their communities. As spokespersons for their neighborhood associations, for example, women gained exposure and experience in dealing with bureaucracies, local officials, and the media. As they gained prominence, these women were invited to serve on city boards and commissions. Similarly, women who volunteered their time to work for a political party or service organization gained experience, useful insights into political processes, administrative skills, and contacts. In some cases, women benefited from mentorship, specific leadership training and development opportunities, and financial or other support when running their own campaigns for public office.

Leaders' Motives for Political Involvement

While these New Mexico leaders were a diverse group with regard to their demographic characteristics and trajectories into politics, they also shared particular goals and reported comparable experiences of political socialization. With few exceptions, these women were not catapulted into politics by some life-altering catalytic event. On the contrary, they were socialized into public life early on by family members and others who helped to prepare them for assuming positions of influence in their communities. Most significantly, these women expressed similar motives for assuming leadership: the desire to fulfill their civic obligation by improving the quality of life for residents of their racial/ethnic communities. In fact, racial and ethnic identity figured prominently in the leaders' motives for political action as they sought to preserve water and soil quality, to revive centuries-old artistic or agropastoral traditions, to repatriate native lands or artifacts, to prevent the commercialization of indigenous cultures, and to politically empower their peoples. Several women were motivated to participate in politics by the goal of preserving traditional New Mexico culture and practices for future generations, despite the tremendous development pressures facing their economically depressed communities.

For the People 147

Characteristic of the environmental justice movement, numerous leaders identified relationships between people of color, their land and labor, and environmental health issues. For example, one leader mobilized against the company that had mined her tribe's land for years, leaving the water and soil severely contaminated. She believed that the industry's mining of uranium on the reservation was directly responsible for the unusually high rate of cancer in the community. A cancer survivor herself, Donna had begun to make connections between patterns she saw in the siting of hazardous facilities and native lands.

> Everywhere you look there are all kinds of Native American Indians . . . dying of cancer. . . . My question is why? Why are these companies trying to store this stuff on our reservations? . . . Are they trying to kill us all? What are they trying to do? . . . We see the scars the company left behind. Why should we let these companies come in and demolish everything . . . in sight? They promise a lot of stuff, but right now I don't think I'd believe any of them anymore. I'm scared about the future for my grandkids. What will there be left for them? (Donna)

A Hispanic leader became politically active after falling seriously ill and being fired from her job at a high-tech manufacturing plant in Albuquerque. Monica's experience motivated her to found an organization with the express purpose of seeking redress in the form of resources and medical services for other women poisoned by their jobs in the computer chip industry. For Monica, inequalities based on gender, race/ethnicity, and class became visible as a result of her own experience; the vast majority of the assembly workers in the industry and, therefore, those most affected by dangerous working conditions, were working-class women of color like herself.

For women of color, the political activities associated with environmental justice are predicated on the critical linkage between race and the environment. Leaders such as Juana saw toxic contamination of their communities as systematic genocide, believing that government and industry target communities of color for environmentally undesirable land uses that will result in the annihilation of their people. Indeed, environmental racism is a recurring theme in many of the narratives of indigenous women activists; as one scholar writes, "it is a genocidal analysis, rooted in the Native American cultural identification, the experience of colonialism, and the imminent endangerment of their culture."[35]

Leaders' Environmental Ideologies

To better understand their conceptualizations of "environmentalism," I examined the basis of the New Mexico leaders' environmental beliefs. In general, these fell within one of four categories: preservation, sustainable use or conservation, environmental justice, and spiritual beliefs. As the leaders' environmental ethics were often richly described and conceptually complex, their beliefs were seldom contained within just one category. Therefore, most of the leaders expressed environmental beliefs that strad-dled, for example, preservation, sustainable use, and environmental jus-tice. Rather than signaling their lack of commitment to one point of view, the findings reveal the multifaceted nature of the leaders' environmental ideologies, their appreciation for the complexities involved in seeking so-lutions to difficult environmental problems, and their understanding of the tremendous cultural variation of perspectives of the environment.

Like mainstream environmentalism, "preservation" stressed the value of maintaining wilderness areas and biodiversity—not only for the enjoyment and use of future generations, but also because nature has value in and of it-self. As one leader stated, an environmentalist is,

> [S]omeone who takes an active role in preserving many things, including cul-ture, nature, people. [Someone] who looks at the world like we all fit together for a reason. If we lose the Mexican wolf or a plant species then we'll lose in-digenous people. There is an interconnectedness. (Lucia)

"Sustainable use" referred to the conservation of land and natural resources and implied a preference for environmentally sustainable forms of economic development. In contrast to preservation, conservation largely placed hu-man needs above those of nonhuman life. To illustrate, one leader was par-ticularly critical of the national environmental movement's lack of concern for issues affecting the lives of poor people and people of color:

> [Mainstream environmentalists] are economically and biologically illiterate; they do not understand local conditions. They take off on their fantasies about a pris-tine type environment and what it should look like and try to apply that to every piece of land. . . . "Trickle down" environmentalism is "If we litigate and if we legislate, it will trickle down and it will create a healthy environment." And just like trickle down economics, it does not get rid of poverty. (Marta)

Support for the principles of environmental justice meant that the leaders' environmental beliefs incorporated issues relative to social and economic justice and included racial and gender equity goals, in addition to improving environmental quality. For example, environmental justice implies sharing both the costs and benefits of environmental policies so that no one racial/ethnic or socioeconomic group assumes disproportionate rewards or burdens (such as the cost of hosting a toxic waste facility). As one activist ex-plained,

> The grassroots environmental justice movement's focus is on survival. Eco-nomic needs must be addressed. The environmental movement is more leisure

and recreation-oriented; they deal with different issues. For example, we need trees for firewood, land for grazing. We see [the environment] as one whole thing. The environment reflects on our spiritual development—protecting the earth for our survival. (Terri)

"Spiritual beliefs," relative to the environment, were grounded in the notion that both the natural environment and all nonhuman life within it are sacred and, therefore, are valued for their own sake. Life was seen as an interconnected web in which maintaining a harmonious balance is essential to its integrity. One woman explained that,

> [Environment] has a different meaning for most Indian people. We can't separate it from our everyday being. You grow up learning and being a part of the environment, not a separate entity. I find it real sad that people separate the words "Indian" and "environmentalist." To me it's one and the same. Like being a civil rights activist and an Indian, they're synonymous. (Carmen)

As indicated in table 8.2, leaders identified themselves as either "environmentalists" or "third world environmentalists." Fewer than half (21, or 42 percent) actually called themselves environmentalists. Most (27, or 54 percent) preferred to identify as "third world" environmentalists, promoting an environmentalism that incorporates participatory decision making and locally determined priorities that reflect the concerns and values of indigenous and third world peoples.

The "environmentalists" demonstrated mainstream environmental values, such as preservation of wilderness areas and maintaining biodiversity, and saw the appropriate role of humans as that of stewards of the earth. These leaders sometimes worked in cooperation with or supported the work of national environmental organizations, approaching policy change from the same ideological perspective: that of classical liberalism. This meant that these self-identified environmentalists believed in the capacity of (and employed) the existing political and economic systems to promote environmental quality within their communities. As one leader stated, "I believe we need to take care of the environment for ourselves and our children. We've seen the benefits of regulations and major legislation such as the Clean Air and Clean Water Acts" (Gina).

Table 8.2. Leaders' Environmental Ideologies

Ideology	Activists Indig.	Hisp.	Officials Indig.	Hisp.	Total	Percent
Environmentalist	4	5	4	8	21	42
Non-Environmentalist	1	0	0	1	2	4
Third World Environmentalist	8	8	9	2	27	54

However, several of the New Mexico leaders objected to the white middle-class bias they believed was inherent in mainstream environmentalism.

> I never said I was an environmentalist, but I believe in the environment. . . . I believe in preserving, in respecting the use of resources. Respect for the environment is essential. If you take care of the earth and you take care of the heavens—important things in life—that will always be there to sustain your life. (Erica)

Despite some of these leaders' involvement in or association with environmental groups, for the most part people of color do not join such organizations for recreational or aesthetic purposes.[36] This is not to suggest that they are not interested in the environment; on the contrary, they may be very concerned with maintaining the health and safety of their communities. Ironically, because racial/ethnic minorities are more likely to experience undesirable or even hazardous living conditions (due to pollutants emanating from nearby industries or waste dumps, for example), they are unlikely to associate their struggle with that of environmentalists. Consequently, twenty-nine of the fifty New Mexico leaders did not consider themselves "environmentalists," even though they espoused values that included respect for nature and preservation of the earth's resources.

The majority of the leaders described themselves as "new," "third world," or "indigenous" environmentalists. As one woman explained,

> An indigenous environmentalist is one who believes in the spiritual value of nature. One who sees nature not only in a patch of forest, but also in the middle of downtown. An indigenous environmentalist does not see anything "wild" about nature. (Linda)

The "third world" environmentalists differentiated their perspective from that of the mainstream environmental movement, claiming that the latter is out of touch with the economic and social realities of the poor, the working class, and people of color. Di Chiro, who interviewed women of color in the environmental movement, similarly found that they were reluctant to call themselves environmentalists "due to the dominance of the mainly white, middle-class, and uncritically 'preservationist' political culture from which much of the mainstream environmental discourse has developed."[37] The principal differences between the New Mexico leaders I interviewed and those in the mainstream environmental movement were their approaches, philosophies, and constituencies. Proponents of the environmental justice movement were concerned with the economic and political empowerment of third world communities. "Environment" was defined broadly to include "where we live, work, and play."

Indeed, the problems that third world environmentalists face are far more immediate and personally threatening than the traditional environmental concern for preservation of wilderness areas and nonhuman species, which they regard as being of less consequence than the survival of their people and their communities.[38] The New Mexico leaders who identified with the environmental justice movement frequently advocated incorporating the cultural beliefs of third world peoples relative to sustainable use (conservation) into contemporary environmentalism. Marta's comments, in particular, illustrate this view.

> When you have settlers that have come onto the land and have survived more than a century, if they came to settle as farmers and ranchers, they will have developed a relationship to the resources in that ecology that they have had to respect [in order to] survive in very harsh climates, which we have. The science that gets learned over several generations, in terms of how people relate to the land and its resources to keep it healthy, is a science that is today being thrown out the window by environmentalists and the government. So keeping the culture strong is an environmental mission. (Marta)

Belief in the value of indigenous people's knowledge and experience of the earth is central to the environmental justice movement. Winona LaDuke, an indigenous environmentalist, ties environmental protection to cultural preservation and economic autonomy for her tribe.[39] The perspectives of the environmental justice movement shared by the third world environmentalists included a strong belief in the rights of citizens to participate in making environmental decisions; a general distrust of government based upon direct experiences with public officials and agencies; a basic belief that human health—rather than aesthetics, wilderness preservation, or other issues—is their primary concern; a skepticism about science and industry; and a prevalent belief that economic growth is not necessarily good and does not benefit everyone equally. Several New Mexico leaders also shared the opinion that "our land is being ripped off with the help of the top ten mainstream environmental organizations that have colluded with industry, the military, and government, and agribusiness" (Juana). These leaders' involvement in environmental issues was often highly personal, because threats to the environment were interpreted as threats to their families and communities.

Leaders' Environmental Policy Initiatives

In the course of their political activism, the New Mexico leaders participated in the formulation and implementation of fifty-one separate environmental policy initiatives. These initiatives incorporated such issues as water quality and water rights, conservation of indigenous or Hispanic peoples' traditional lands and land uses, long-range land-use planning, preservation of soil quality, waste management, and production and use of renewable energy sources. Specifically, grassroots activists pressed for policies to provide information and technical assistance to communities of color for the purpose of developing environmental programs; facilitate the negotiation of agreements between communities and industry or government agencies for pollution abatement and site cleanup; encourage increased citizen participation in environmental policymaking; create programs for the prevention of soil and groundwater contamination; and provide reparation to victims of toxic poisoning. Public policies advocated by officials included similar components, such as the development of regulations and implementation of programs to enhance air and water quality; the cleanup of toxic sites; ensuring tribal control of mineral rights on Indian lands; and the formulation of land-use plans to guide urban development. As one leader stated,

> For several years, I've sponsored a beverage container recycling act—the "bottle bill." I carry legislation that relates to sewer and water hookups for low-income people so that we can get them into systems that will give them safer water. I like issues that relate to landfills, reduce and reuse of materials, safety for our water table. Mostly anything that protects our water and air, our infrastructure. (Sonia)

As is characteristic of the environmental justice movement, the leaders' policy agendas included human rights and public service initiatives, which were often tied to more "traditional" environmental concerns, such as sustainable development. For example, promoting sustainable economic development, preserving traditional agricultural practices, and improving the Rio Grande's water quality were equally important goals for one leader who stated that her organization's mission was "to restore the water to drinkable quality and to keep the communities active and alive along the Rio Grande" (Miranda). Several of the leaders believed that communities of color are specifically targeted for placement of polluting industries and facilities. To support their claims, they cited specific examples of cancer clusters, the contamination of well sites, incidences of health and safety violations by both government and industry, and other such problems. Their efforts to publicize these concerns resulted in the production of video documentaries aired by local public television stations, the publication of numerous studies and position papers by various organizations, public hearings held around the state, and feature stories in the news media.

The process of formulating public policy requires that (democratic) governments provide the means by which citizens, both as individuals and as representatives of groups, are enabled to participate actively. The New Mexico leaders recognized that community members must be included as principal actors in the policy-making process to identify local needs and to permit consultation prior to policy design, during program implementation, and throughout evaluation procedures. Empowering community members to effectively participate in public decision making was a policy priority for both activists and of-

ficials. The New Mexico leaders promoted consensus building, shared decision making, and community participation in goal setting and public policy
formulation. Each leader exhibited egalitarian views and supported participatory forms of democracy. Some strove for political and economic transformation of their racial/ethnic community by increasing public awareness and access to information through forums, workshops, and public hearings. Others
sought to empower community members through leadership training and education. In several cases, grassroots activists recruited their membership to
promote gender equality and ethnic/racial diversity and to ensure representation of poor and low-income citizens in their organizations' leadership. In general, the New Mexico leaders sought to increase political participation by en-
abling community members to help themselves, by providing advocacy, by
facilitating coalition building, and by sharing information.

My findings suggest that the fifty New Mexico leaders are devoted political actors who value democratic participation and work to improve social,
environmental, and economic conditions in their communities. Their participation is animated by the overarching goal of achieving economic equity
and social justice for their communities. The politics of these leaders are
heavily influenced by their racial/ethnic and gender identity as they seek to
represent the interests of their communities, to advance the position of
women in society, and to empower individual citizens to participate actively
in public life. Both their racial/ethnic identity and environmental ideology inform and shape their public policy agendas and their roles in New Mexico's
environmental politics.

CONCLUSION

The New Mexico leaders' environmental ideologies and public policy initiatives reflected, to varying degrees, the history and situation of their
racial/ethnic group in the United States, the leaders' motives for political involvement, identification with their racial/ethnic group, and personal experience of politics. In general, they rejected the mainstream Euro-American
environmental movement in favor of alternative ideologies that incorporated
the cultural values, beliefs, and life situations of people of color into their political agendas. Together, the motives, political goals, and ideology of the
New Mexico leaders shaped their public policy agendas.

While their agendas included a broad array of concerns, they generally
focused on three shared objectives: (1) to safeguard the rights of community members and empower them to participate in public decision-making
processes, (2) to secure services and resources to improve social and economic conditions for racial/ethnic minorities and for women, and (3) to
promote the conservation of natural resources and the use of sustainable

development practices. Although the scope of their work extended from solving neighborhood-level problems to forming networks with organizations internationally, their common focus was achieving environmental justice for their racial/ethnic communities. This entailed providing for the health and safety of community members, empowering politically marginalized individuals and groups, and establishing programs to meet their social and economic needs. While the public officials advocated more policies aimed at community development and the grassroots activists pursued greater numbers of environmental policies, the New Mexico leaders overwhelmingly supported policies to empower community members.

Significantly, whether they were activists or officials, indigenous or Hispanic, the leaders' policy initiatives reflected the goals of the environmental justice movement, incorporating such notions as gender equity; social, economic, and environmental justice; and increased participation in public decision making. These research findings are significant for at least a couple of reasons. First, they demonstrate greater support for "mainstream" environmentalism by American Indian and Hispanic women than might be expected based on existing literature. Second, they reveal the ingenuity of the New Mexico leaders in adapting elements of mainstream social movements to the needs of their own particular racial/ethnic communities. Furthermore, the alternative forms of environmentalism espoused by the leaders reveal their underlying concerns with preserving their culture, achieving racial and gender equity, increasing political participation, and maintaining or improving environmental quality. These findings are also important because while they identify common themes and confirm the results of a handful of studies that examine the political goals and public policy agendas of American Indian and Hispanic women.[40] In combination, these studies reveal that indigenous and Hispanic women leaders enter politics for the people—in order to improve the quality of life in their racial/ethnic communities by empowering community members and increasing participation in public decision-making processes.

NOTES

1. See especially S. Rodriguez, "Land, Water, and Ethnic Identity in Taos," in *Land, Water, and Culture: New Perspectives on Hispanic Land Grants,* ed. C. L. Briggs and J. R. Van Ness, 313–403 (Albuquerque: University of New Mexico Press, 1987).

2. M. Guerrero, "Albuquerque Revisits Radioactive Dumping," *Voces Unidas* 4, no. 2 (1994): 5; M. L. Jones, "Missiles over Dineh," *Voces Unidas* 4, no. 1 (1994): 11; and C. Seidman, "That Their Children May Inherit the Earth," *Sage Magazine* (October 1994): 10–12.

3. D. Salazar and L. A. Moulds, "Toward an Integrated Politics of Social Justice and Environment: African American Leaders in Seattle" (paper presented at the annual

meeting of the Western Political Science Association, Portland, Oregon, March 1995);
B. Marquez, "The Politics of Environmental Justice in Mexican American Neighborhoods" (paper presented at the annual meeting of the Western Political Science Association, Albuquerque, New Mexico, March 1994); and L. Head, "Take Back New
Mexico!" *Voces Unidas* 3, no. 2 (1993): 1 (2).

4. J. Fleck, "Nuclear Waste Plan Rejected," *Albuquerque Journal,* February 2,
1995; Partners in the Environment, "New Mexico: The Land of Enchantment?" *EAGLE*
1, no. 6 (1993): 5 (2).

5. R. Contreras and R. Shaw, "Isleta Pueblo's First Woman Governor," *Voces
Unidas* 3, no. 2 (1993): 10 (2), and T. Anaya and B. F. Chavis Jr., "People of Color
Unite to Combat Environmental Racism," *Albuquerque Journal,* October 25, 1991.

6. T. Enslin, "Keep Radioactive Waste Water Out of Sewers," *Albuquerque Journal,* November 1, 1991; F. Moreno, producer, *Environment, Race and Class: The Poisoning of Communities of Color* (Albuquerque, N. Mex.: Community Cable Channel
27), aired August 31, 1991.

7. V. Taliman, "Uranium Miners Made Profits, Left Town, and Stuck Natives with
Open Pits and Contamination Worries," *Voces Unidas* 1, no. 3 (1991): 6.

8. Inter-Hemispheric Education Resource Center, "Concerned Citizens of Sunland
Park," *BorderLines* 2, no. 1 (1994): 4–5, and R. Fernandez, "Los Lunas Stops Waste Incinerator," *Voces Unidas* 3, no. 3 (1993): 12.

9. L. Naranjo, "Martineztown Defeats Courthouse!" *Voces Unidas* 4, no. 2 (1994): 4.

10. Jones, "Missiles over Dineh"; A. Gedicks, *The New Resource Wars: Native and
Environmental Struggles against Multinational Corporations* (Boston: South End,
1993); and SouthWest Organizing Project, "Carletta Tilousi: Fighting for Her People,"
Voces Unidas 1, no. 4 (1991): 13.

11. L. Pulido, "Sustainable Development at Ganados del Valle," in *Confronting Environmental Racism: Voices from the Grassroots,* ed. R. D. Bullard (Boston: South
End, 1993), 123–39, and D. Peña, "The 'Brown' and the 'Green:' Chicanos and Environmental Politics in the Upper Rio Grande," *Capitalism Nature Socialism: A Journal of Socialist Ecology* 3, no. 1 (1992): 79–103.

12. R. M. Laws, "Spiritual Conflict Underlies Nuclear Waste Controversy," *Albuquerque Journal,* March 8, 1995, and L. Taylor et al., "The Importance of Cross-Cultural Communication between Environmentalists and Land-Based People," *Workbook* 13, no. 3 (1988): 90–93.

13. Seidman, "Inherit the Earth"; and R. Contreras, "Toxic Survivors Seek Protection," *Voces Unidas* 3, no. 2 (1993): 7.

14. F. Romero, "Grupitos: Citizen Involvement at the Kitchen Table," *Nuestro
Pueblo* 7, no. 2 (Summer 1997); SouthWest Organizing Project, "Five Years Later:
SWOP's Letter to the 'Group of Ten' Revisited," *Voces Unidas* 5, no. 3 (1995), special
insert: 1–4; and SouthWest Organizing Project, "The Selling of New Mexico," *Voces
Unidas* 4, no. 2 (1994a), special insert.

15. D. M. Prindeville and J. G. Bretting, "Indigenous Women Activists and Political
Participation: The Case of Environmental Justice," *Women & Politics* 19, no. 1 (1998):
39–58; C. Mack-Canty, "Women Political Activists: Taking the State Back Out" (paper
presented at the annual meeting of the Western Political Science Association, Tucson,
Arizona, March 1997); and SouthWest Organizing Project, "Grassroots Democracy in
Action: An Interview with Richard Moore," *Voces Unidas* 4, no. 3 (1994b): 9 (3).

16. See Salazar and Moulds, "Toward an Integrated Politics"; D. Alston, *"Taking Back Our Lives"*: *A Report to the Panos Institute on Environment, Community Development and Race in the United States* (Washington, D.C.: Panos Institute, 1990), and M. Guerrero and L. Head, "Informed Collective Action—A Powerful Weapon," in *Taking Back Our Lives,* ed. Alston, 24 (4).

17. Marquez, "Politics of Environmental Justice."

18. Prindeville and Bretting, "Indigenous Women Activists"; S. Cable and C. Cable, *Environmental Problems, Grassroots Solutions: The Politics of Grassroots Environmental Conflict* (New York: St. Martin's, 1995), and D. Taylor, "Environmentalism and the Politics of Inclusion," in *Confronting Environmental Racism,* ed. Bullard, 53–61.

19. N. E. McGlen and K. O'Connor, *Women, Politics, and American Society* (Englewood Cliffs, N.J.: Prentice Hall, 1995), and J. M. Bystydzienski, "Introduction," in *Women Transforming Politics: Worldwide Strategies for Empowerment,* ed. J. M. Bystydzienski, 1–8 (Indianapolis: Indiana University Press, 1992).

20. T. Aragón de Valdez, "Organizing as a Political Tool for the Chicana," *Frontiers* 5, no. 2 (1980): 7.

21. L. A. Tilly and P. Gurin, "Women, Politics and Change," in *Women, Politics and Change,* ed. L. A. Tilly and P. Gurin, 3–34 (New York: Russell Sage Foundation, 1990), and S. Morgen and A. Bookman, "Rethinking Women and Politics: An Introductory Essay," in *Women and the Politics of Empowerment,* ed. A. Bookman and S. Morgen, 3–29 (Philadelphia: Temple University Press, 1988).

22. C. Hardy-Fanta, *Latina Politics, Latino Politics: Gender, Culture, and Political Participation in Boston* (Philadelphia: Temple University Press, 1993); S. Cable, "Women's Social Movement Involvement: The Role of Structural Availability in Recruitment and Participation Processes," *Sociological Quarterly* 33, no. 1 (1992): 35–50; and L. Albrecht and R. M. Brewer, "Bridges of Power: Women's Multicultural Alliances for Social Change," in *Bridges of Power: Women's Multicultural Alliances,* ed. L. Albrecht and R. M. Brewer, 2–22 (Philadelphia: New Society, 1990).

23. N. A. Naples, "Women's Community Activism: Exploring the Dynamics of Politicization and Diversity," in *Community Activism and Feminist Politics: Organizing across Race, Class, and Gender,* ed. N. A. Naples, 327–49 (New York: Routledge, 1998), and N. A. Naples, "Activist Mothering: Cross-Generational Continuity in the Community Work of Women from Low-Income Urban Neighborhoods," *Gender and Society* 6, no. 3 (1992): 441–63. Also see A. Orleck, "Tradition Unbound: Radical Mothers in International Perspective," in *The Politics of Motherhood: Activist Voices from Left to Right,* ed. A. Jetter, A. Orleck, and D. Taylor, 3–11 (Hanover, N.H.: University Press of New England, 1997).

24. McGlen and O'Connor, *Women, Politics;* S. Thomas, *How Women Legislate* (New York: Oxford University Press, 1994); and Hardy-Fanta, *Latina Politics.*

25. G. Di Chiro, "Defining Environmental Justice: Women's Voices and Grassroots Politics," *Socialist Review* 22, no. 4 (1992): 109.

26. Di Chiro, "Defining Environmental Justice," 109. Also see R. Paehlke and P. Vaillancourt Rosneau, "Environment/Equity: Tensions in North American Politics," *Policy Studies Journal* 21, no. 4 (1993): 672–86.

27. Mack-Canty, "Women Political Activists"; Taylor, "Environmentalism"; and Cable, "Women's Social Movement Involvement."

28. T. Kaplan, *Crazy for Democracy: Women in Grassroots Movements* (New York: Routledge, 1997), and Seidman, "Inherit the Earth."

29. Di Chiro, "Defining Environmental Justice," 115.

30. U.S. Census Bureau, *Census 2000 Summary File 2,* Matrices PCT3 and PCT4. "Census of Population—Geographic Area, New Mexico Race, or Ethnic Groups." <http://factfinder.census.gov/servlet/QTTable?ds_name=DEC_2000_SF2_U&geo_id =04000US35&qr_name=DEC_2000_SF2_U_QTP1> (May 6, 2002).

31. J. Lofland and L. Lofland, *Analyzing Social Settings: A Guide to Qualitative Observation and Analysis,* 3d ed. (Belmont, Calif.: Wadsworth, 1995).

32. M. S. Feldman, *Strategies for Interpreting Qualitative Data* (London: Sage, 1995), and G. McCraken, *The Long Interview* (Newbury Park, Calif.: Sage, 1988).

33. Feldman, *Strategies;* Lofland and Lofland, *Analyzing Social Settings;* and M. B. Miles and A. M. Huberman, *Qualitative Data Analysis,* 2d ed. (London: Sage, 1994).

34. J. A. Holstein and J. F. Gubrium, *The Active Interview* (Beverly Hills, Calif.: Sage, 1995); Lofland and Lofland, *Analyzing Social Settings;* and A. Strauss and J. Corbin, *Basics of Qualitative Research: Grounded Theory Procedures and Techniques* (London: Sage, 1990).

35. C. Krauss, "Women and Toxic Waste Protests: Race, Class and Gender as Resources of Resistance," *Qualitative Sociology* 16, no. 3 (1993): 257. Also see W. Churchill and W. LaDuke, "The Earth Is Our Mother: Struggles for American Indian Land and Liberation in the Contemporary United States," in *The State of Native America: Genocide, Colonization, and Resistance,* ed. M. A. Jaimes, 139–88 (Boston: South End, 1992), and M. A. Jaimes, "American Indian Women: At the Center of Indigenous Resistance in North America," in *The State of Native America,* ed. Jaimes, 311–44.

36. Taylor, "Environmentalism," and Alston, "Taking Back Our Lives."

37. Di Chiro, "Defining Environmental Justice," 94.

38. Marquez, "Politics of Environmental Justice."

39. Orleck, "Tradition Unbound."

40. See for example, M. Pardo, "Doing It for the Kids: Mexican American Activists, Border Feminists?" in *Feminist Organizations: Harvest of the Women's Movement,* ed. M. M. Ferree and P. Y. Martin, 356–71 (Philadelphia: Temple University Press, 1995); K. B. Chiste, "Aboriginal Women and Self-government: Challenging Leviathan," *American Indian Culture and Research Journal* 18, no. 3 (1994): 19–43; P. Cruz Takash, "Breaking Barriers to Representation: Chicana/Latina Elected Officials in California," *Urban Anthropology* 22, nos. 3–4 (1993): 325–60; and M. McCoy, "Gender or Ethnicity: What Makes a Difference? A Study of Women Tribal Leaders," *Women & Politics* 12, no. 3 (1992): 57–68.

IV

PROSPECTS FOR THE FUTURE

9

In Pursuit of Healthy and Livable Communities

Estelle Bogdonoff, Kathleen Cooper-McDermott, and Kenny Foscue

An ecological approach to public health examines the environmental hazards and environmental justice concerns that affect the health of a population or group. Environmental hazards frequently impact a specific community or ethnic/racial group unfairly, based on where they live or the nature of their housing. Low socioeconomic status populations are considered to be the most vulnerable and often at higher than average risk of disease. When physical environments are unhealthy, both individual and community health are affected. Yet the efforts of these populations to prevent environmental damage to their communities or to improve their environment may be hindered by a lack of information or of political representation or influence.

Public policy is integral to the environment, shaping the conditions for health and creating options from which diverse social groups choose their lifestyles.[1] Governmental policies provide resources, distributed by state and local organizations, that shape home, school, and community environments. This in turn affects the patterns of personal and community living and creates the potential for individual health and welfare. As awareness of environmental health risks has increased, the public has called for more response by the public health sector. In Connecticut, illustrations of approaches to address environmental health hazards can be found at the federal, state, regional and local levels. This chapter spotlights instances of proactive and supportive public policy and the narrowing of health disparities. The following sections discuss successful collaboration by state, federal, regional, and local agencies and organizations.

- Section 1. In "Public Health and Brownfields," Kenny Foscue, Division of Environmental Epidemiology and Occupational Health, Connecticut

Chapter 9

Department of Public Health, describes the need to address real and perceived public health issues at Brownfields sites and shows how converting these potentially hazardous sites to healthy space can benefit communities when public health and economic interests work together.

- Section 2. In "Connecticut Agencies and Organizations Develop a Coordinated Program to Promote EPA Tools for Schools," Kenny Foscue describes the work of the Connecticut School Indoor Environment Resource Team. This collaborative approach is used to address school indoor air quality as a major public issue.
- Section 3. In "Coalition Building at the Local Level," Estelle Bogdonoff, cochair, Southeastern Connecticut Indoor Air Quality Coalition, examines the need and principles of building a county coalition to address indoor air quality problems in the home, school, and workplace.
- Section 4. In "People, Places and Asthma: Its Ecological Imprint in Our Midst," Kathleen Cooper-McDermott, New London Department of Health and Social Services, discusses a program developed by a small city health department working with the local hospital to reduce the burden of asthma among city residents and decrease the number of visits to the emergency department and admissions to the hospital.

PUBLIC HEALTH AND BROWNFIELDS

While Brownfields site activities are largely driven by economic and environmental considerations, there is a strong need for public health involvement. As municipalities move forward with Brownfields initiatives, it is imperative that public health issues are considered from the earliest phases of site selection, evaluation, and development. Furthermore, it is necessary that all Brownfields stakeholders work in a coordinated manner. In general, experience indicates that early, consistent public health involvement at Brownfields and other hazardous waste sites results in more of the effective and appropriate addressing of public health concerns, better communication and involvement with communities, and more timely and efficient cleanup and development of sites.

What Are Brownfields?

The U.S. Environmental Protection Agency (EPA) defines Brownfields as "abandoned or idle industrial and commercial facilities generally in urban areas, where redevelopment is complicated by real or perceived contamination."[2] A multiagency initiative in conjunction with the private sector has been in existence for the past several years to redevelop these sites. The EPA has been the lead agency, offering several grant programs that seek to facilitate Brownfields development on a local level. The Brownfields Showcase Communities Program is the largest of these initiatives. Stamford, Connecticut, has received funding under this national program. Another EPA Brownfields program is the Regional Assessment Demonstration Pilot. These are primarily funds provided to local entities for site testing for single sites in designated cities. The third funding initiative is the National Assessment Demonstration Pilot program, which awards grants of approximately $200,000 to city governments to assess site contamination. An environmental justice and community outreach component is usually a requirement for funding.

The Connecticut Department of Environmental Protection (DEP) has a state-level Brownfields initiative entitled the Urban Sites Remediation Action Program. This effort provides funding and staff assistance to Connecticut cities and towns to address urban sites. A principal component of this program entails the use of licensed environmental professionals (LEPs) to conduct assessments on Brownfields sites.

Interrelationship between Brownfields and Environmental Justice

There is a strong interrelationship between the concept of Brownfields and environmental justice, defined by the EPA as the "fair treatment for people of all races, cultures, and incomes, regarding the development of environmental laws, regulations, and policies."[3] A growing recognition has developed that minority and low-income populations bear a disproportionate burden of adverse health and environmental effects. Research has established a relationship between proximity to hazardous waste sites and race/income.[4] Many of these sites are in urban areas and are officially designated Brownfields sites. Charles Lee succinctly describes this relationship between Brownfields and environmental justice in *Environmental Justice, Urban Revitalization and Brownfields*:

> The Brownfields issue is yet another aspect of an intensifying set of systemic problems related to residential segregation, disinvestment of inner-city areas, urban sprawl, degradation of the urban environment, and the polarization between urban and non-urban communities along lines of age, life style, race, socioeconomic status, and other spatially-related social divisions. These are endemic to a severe crisis—environmentally, economically, socially, culturally, and otherwise—in urban America. Environmental justice encompasses very clearly the inextricable linkage between these issues.[5]

Brownfields Redevelopment Can Improve Public Health

In the broadest view, Brownfields redevelopment can potentially improve the overall health of the community by bringing in jobs to economically

164   Chapter 9

depressed areas, lowering poverty levels, and increasing access to health care. An indirect outcome may be a reduction in the crime rate. A more direct health benefit is the remediation of hazardous sites that may pose short- and long-term health risks to the community. These abandoned sites are attractive to children who may be injured or exposed to toxic chemicals. In addition, there can be an overall benefit to the "mental health" of the community. Revitalizing these sites may help neighborhoods gain a more positive self-image. Residents may feel more "invested" in their neighborhoods, which may contribute to further financial investment. Another factor is research data that document the relationship between higher levels of stress and proximity to hazardous waste sites. The redevelopment of Brownfields sites may have a positive impact by lowering community stress.

Public Health Is an Important and Necessary Partner in Brownfields Initiatives

At heavily contaminated sites, it is important to evaluate health risks based on exposure pathway analysis. Exposure pathway analysis links contaminant sources, locations, and environmental releases with population activities and locations. People may be exposed in ways that may not be identified by only reviewing environmental data. Exposure pathway analysis by qualified public health professionals may identify these potential exposures. Public health agencies can address health concerns and real or perceived health risks earlier in the process to prevent problems that may slow down or prevent redeveloping a site. Public health involvement may also include a review of site cleanup plans by state environmental health specialists, to ensure that the health of the surrounding neighborhood and of the workers conducting the cleanup are protected. In addition to protecting the public health from potential exposures, this oversight may contribute to community acceptance of remediation and development plans and activities.

Another important role of all public health agencies is to ensure community participation in the Brownfields redevelopment process. The surrounding neighborhood residents and businesses should be informed and allowed access to and participation in decision making. Public health agencies, particularly local health representatives, have an important role in making sure the community's health concerns are adequately addressed. Answering potential or perceived concerns early on may prevent development conflict slowdowns later. For example, the residents around a Brownfields site may have past knowledge of site activities that involved the use of hazardous chemicals. The residents may fear that site remediation activities will lead to exposure. Public health agencies, with assessment capabilities, may be able to answer these concerns (including providing recommendations to prevent exposures), allowing the remediation to continue on schedule, without potential opposition from the community. Local health departments have a unique role, as they are in closer proximity and usually have more ties to the local community.

Public Health Questions at Brownfields Sites

In remediating Brownfields sites, key public health questions must be asked in investigating and evaluating a site. Foremost, what is the future use of the site: residential? daycare? office? These types of uses could require more site investigation and a more stringent site cleanup protocol. Another question arises regarding exposure during remediation. Even if the environmental site assessment does not find contaminants above the regulation levels, this does not mean all health concerns have been answered. An example of this may be the need for indoor air sampling in a former industrial building that is being considered for residential use. Site cleanup plans may require outdoor air sampling to check for potential exposures to the neighborhood or to site workers. In any case, it is important that health-based sampling protocols be utilized, with oversight from public health agencies. These agencies, especially local health departments, can develop strategies to answer the community's health concerns (i.e., What was the potential for past exposures? Are there indoor air quality and other indoor environmental issues to be addressed? How will the neighborhood be protected during remediation?) These agencies may be viewed as more objective and therefore more trustworthy in addressing community concerns.

How Much Public Health Involvement at Brownfields Sites?

On most Brownfields sites, there is little or no need for public health agency involvement. Presently, there is no official legislation or protocol that prescribes when health agencies should or can review Brownfields site data in an effort to identify and resolve potential public health issues. There are, however, a number of priority sites that should trigger public health involvement. These include:

- sites where future development includes residential land use;
- sites where development would include significant contact by sensitive/ vulnerable populations, particularly children (i.e., schools, day-care facilities, and recreational areas);
- sites with significant contamination located in or adjacent to residential areas;
- sites slated for high-use office space; and
- sites slated for industrial or commercial use but known or suspected to be highly contaminated.

CONNECTICUT AGENCIES AND ORGANIZATIONS DEVELOP A COORDINATED PROGRAM TO PROMOTE THE ENVIRONMENTAL PROTECTION AGENCY'S "TOOLS FOR SCHOOLS"

Of increasing concern to public health are the environmental hazards to which people are exposed through indoor air. Children, a high-risk group, spend much of their day inside school; therefore, consideration of the indoor air quality of schools is significant. The EPA has taken a lead in confronting this issue by developing the Tools for Schools (TFS) program. In Connecticut, several agencies and organizations concerned with school indoor air quality have banded together to promote the TFS program as a means of addressing indoor air quality and assisting schools in implementing the program. After two years of collaborative work, this strategy is paying off. The Connecticut School Indoor Environment Resource Team has conducted extensive outreach to Connecticut schools, developed and implemented a comprehensive training program assisting more than seventy schools to establish TFS committees, and encouraged state policy makers to address school indoor air quality (IAQ) as a major public health issue.

What is "Tools for Schools"?

The EPA developed an innovative program—Indoor Air Quality Tools for Schools Action Kit—to enable schools to identify and address IAQ problems. The program is based on three key principles:

• The school community can prevent many IAQ problems.
• IAQ problems can often be resolved using the skills of school staff.
• The expenditures and effort to prevent most IAQ problems are a fraction of that required to solve problems once they develop.

The TFS action kit provides all the materials necessary to promote a low-cost, problem-solving team approach to improving IAQ. Once a committee of administrators, teachers, maintenance staff, parents, and others investigates and prioritizes indoor air hazards, short- and long-term strategies are developed to solve IAQ problems.

Background

Like many states, Connecticut has a hodgepodge of agencies and organizations that have some responsibility for school IAQ. Connecticut has a state Occupational Safety and Health Administration (OSHA) program that covers municipal and state worker occupational issues, but like its federal counterpart it is principally standard driven. Because there are no IAQ standards, the agency is limited in its response to school employee IAQ complaints, though it does provide inspection, testing, and consultation services to schools. The state health department—the Connecticut Department of Public Health (CT DPH)—does not have a formal IAQ program. The Division of Environmental Epidemiology and Occupational Health does respond regularly to general and school IAQ complaints, principally by providing phone consultation, education, and referral services. The occupational and environmental medicine programs at the University of Connecticut and Yale University regularly examine teachers and other school personnel regarding indoor air exposures and provide industrial hygiene services. The Connecticut Department of Education has had very limited involvement in school IAQ issues until recently. The Connecticut Department of Environmental Protection (DEP) has an urban environmental program with some history of involvement with urban schools, primarily in Hartford. The Connecticut Council for Occupational Safety and Health (CTCOSH) is a labor/health professional-based advocacy organization that has had a small grant from EPA to promote the TFS program for several years. Recently, the American Lung Association (ALA) of Connecticut has received funding for TFS activities and has become very active in TFS activities in Connecticut.

Several of these agencies, along with staff from the EPA New England regional office, made efforts in the past to promote the TFS program in Connecticut, with limited success. Over the years, EPA distributed the kit to most schools in Connecticut, but provided little organized follow-up. CTCOSH assisted some parents and teachers in Hartford and New Haven in initiating TFS committees, but with little or no support from the school systems, these committees were largely unsuccessful. CT DPH worked with a few local health directors to encourage problem schools to implement the program, but with no immediate success.

A New Approach

In January 1999, staff from CT DPH, CTCOSH, and EPA decided to pool resources and develop a coordinated response to school IAQ. The goal of this collaboration was to develop a team of professionals from various agencies and disciplines who would promote TFS, develop a training program for TFS school committees, and provide ongoing technical assistance. The Connecticut School Indoor Environment Resource Team has grown to include members who come from federal, state, and municipal agencies and from labor, health advocacy, and community organizations: CT DPH, the Connecticut Department of Education, the Connecticut DEP, CTCOSH, EPA New England, the University of Connecticut Division of Occupational and Environmental Medicine, the Connecticut Association of Local Health Directors, the Connecticut

School Building and Grounds Association, ALA of Connecticut, the Connecticut Education Association (CEA), the Connecticut Federation of Educational and Professional Employees, the Civil Rights Division of the U.S. Department of Education, the Connecticut Association of Boards of Education, the Yale Occupational and Environmental Medicine Program, and the Southeastern Connecticut Indoor Air Quality Coalition.

It is important to recognize the active participation of the Connecticut Education Association, which since its early involvement has contributed substantial time and resources. The Resource Team includes IAQ specialists, industrial hygienists, epidemiologists, educators, trainers, and others who can assist schools in implementing the TFS program. The Resource Team provides the following services:

- outreach and education through meetings and presentations with school officials and other stakeholders (e.g., teachers, PTA) to promote TFS in school systems;
- a two-session training program for TFS committees that includes IAQ health issues, how TFS works, communication, conducting site walk-throughs, prioritizing IAQ problems; and
- ongoing consultation with TFS committees to set priorities and address specific technical questions.

Overcoming Roadblocks and Barriers

The Resource Team faced several roadblocks and barriers to the successful implementation of the program by local school systems.

Getting to "First Base": How to Encourage Schools to Consider Implementing TFS

The original members of the Resource Team—staff from CTCOSH, CT DPH, the occupational medicine programs, and EPA—individually and as a group had conducted outreach and education efforts to several schools, teachers' union locals, PTA meetings, and other school-related groups. Additionally, DPH staff had been working with a few local health departments. Although there was strong interest among many of these groups, school systems were reluctant to embrace the program. Reasons included a general fear about addressing an unknown issue, a perceived lack of expertise regarding IAQ, fear about the time commitment required to implement and maintain the program, an unwillingness to address what was perceived as a problem of "a few complainers," and a fear of "opening Pandora's Box." The group decided that establishing a team of experts who could assist schools to implement the program might make the TFS program more attractive. This included the development of a training program to educate the TFS committees about IAQ issues and how to use the kit. In addition, members of the Resource Team would be available for advice and consultation as the committees implemented the program. After the training program was established, the Resource Team publicized its services through mailings to public school superintendents from the state health and education commissioners, through group meetings, and through limited media coverage.

Getting School Systems to Endorse and Adopt the Program in a More Timely Manner

Because the long-range goal has been to encourage and support TFS implementation in as many schools as possible, the Resource Team has evolved strategies to attain that goal. In early recruitment efforts, significant time and energy was spent on meetings with individual school principals, union locals, and other local school community stakeholders. In addition, the Resource Team conducted presentations at several regional and statewide organizations, such as the two teachers unions, the Connecticut Education Association (CEA) and the state American Federation of Teachers (AFT) affiliate, the statewide school nurses association, and the Connecticut Association of Boards of Education (CABE). The two teachers unions put considerable effort into contacting their locals and individual members and encouraging them to promote the program in their local schools. Although all of these efforts have been useful, the Resource Team has limited resources and time. For this reason, it began to focus on strategies that were more time efficient and successful at getting schools—and particularly school systems—to commit to adopting the program. In reviewing the process used, it was determined that the most effective procedure was to meet with a school system's superintendent and gain his/her endorsement. School systems are generally a traditional hierarchy, with the superintendent making such decisions. The first major effort involved conducting a statewide mailing to all superintendents, followed by a telephone call. This resulted in several meetings, which led to implementation in several schools.

However, another problem—making sure the whole school system had "bought into" the TFS program—needed to be addressed. It became apparent that key personnel in the school need to be on board and knowledgeable about TFS. The Resource Team's current protocol is to meet with and present information on TFS to a school system's executive staff—the superintendent, principals, the facilities manager, the nurse coordinator, and preferably the president of the local teacher's union chapter. The presentation stresses why it is in the school system's interest to implement TFS. Special care is taken to get the facilities manager on board and to get a commitment from the school system to adopt the program.

Encouraging School Systems to Adopt the Program System-wide

During the early stages of the Resource Team's outreach, school systems were encouraged to "experiment" with the program in one or two schools, particularly if the administration appeared wary. As more and more schools in Connecticut began to adopt the program, the Resource Team became overwhelmed with trainings. It became clear that the goal should be to require school systems to adopt TFS system-wide or in several schools at a time. Furthermore, training sessions would become more effective as committee members from the different schools could compare notes on their experiences.

Methods

The Resource Team uses six methods to promote TFS and the use of the Resource Team itself. These are:

- Materials development and distribution. This includes brochures, fact sheets, memos, and training resources.
- Meeting presentations. A key strategy to encourage school systems to adopt TFS is to make presentations to the executive staffs of local school systems or school system–wide health and safety committees. In addition, the Resource Team makes presentations to local and statewide agencies and organizations to seek their help in publicizing and implementing TFS.
- Mailings. Member organizations of the Resource Team have conducted several mailings to the targeted audience, including a mass mailing to all Connecticut superintendents.
- Conferences. The Resource Team has participated in several conferences related to school IAQ issues. In addition, the Resource Team sponsored two highly successful conferences: a regional in-service training conference for Hartford area school nurses (March 14, 2001) and a statewide conference, "Indoor Air Quality for Connecticut's Schools: Toward Healthy and Safe School Environments" (October 16, 2001).
- Newsletters, newspapers, and other media. Several newspaper and newsletter articles have been written about school IAQ, EPA's TFS, and the Resource Team.
- Networking with Resource Team member organizations. Resource Team member agencies and organizations have made continuing efforts to mobilize their members to assist our efforts. This includes the teachers unions, the ALA, the Connecticut Association of Local Health Directors, and CABE.

The TFS Training Program

A primary activity of the Resource Team has been the development and implementation of a two-step training program for TFS committees. By providing extensive training, committees are more likely to be effective in implementing TFS. The two training sessions are conducted in a workshop format, with emphasis on participatory learning techniques such as group exercises. Before the first workshop is conducted, the TFS kit and other information are mailed to each school team coordinator. The first session is a three-hour workshop that covers an overview of school IAQ issues, an exercise and presentation on IAQ health issues, and a detailed review of how to use TFS. Two or three trainers from the Resource Team conduct the workshop.

The second training session is a two-hour workshop on conducting walk-through investigations. This workshop is conducted by one of the industrial hygienists from the Yale or University of Connecticut Environmental Medicine programs or the Connecticut OSHA program. At least one of the first session trainers also facilitates this session. The basics of building IAQ investigations are covered and a group investigation exercise is conducted in one or two areas of the training site (a school).

Accomplishments

From its inception in 1999 through October 2001, the Connecticut School Indoor Environment Resource Team has cataloged an extensive list of successes. The Resource Team has assisted in the implementation of TFS committees in more than seventy-five schools in twenty-five Connecticut school systems and has conducted more than forty training workshops for TFS committees. This has been accomplished by an aggressive outreach strategy and the successful collaboration of many organizations and agencies. In addition, several of the Resource Team member organizations have taken important roles in promoting a school IAQ bill in the state legislature.

COALITION BUILDING AT THE LOCAL LEVEL

The conjoining of resources to address public health issues is increasingly becoming the "modus operandi" at both the state and local levels. While collaboration is not a new phenomenon, what is new is that these collaborations are becoming more formally recognized. Moreover, these collaborations extend beyond merely sharing resources. Communities are taking it upon themselves not only to focus on the problem at hand, but also to develop the skills needed to sustain its efforts over the long term, rather than merely completing a specific project and disbanding.

Coalition-Building Phases

A coalition is defined as individuals or organizations working together in a common effort for a common purpose to make more effective and efficient use of resources—an alliance.[6] Whereas a network is a gathering of individuals or organizations that share information, ideas, or resources, a coalition implies united action(s), allowing diverse groups to come together in a common cause. Building a coalition occurs in steps and the speed in which it is built depends on the urgency of the situation and on the active commitment of the group. The first three phases of coalition building are (1) forming a core initiating group to lead the effort; (2) completing a stakeholder analysis, including developing recruitment strategies, and finalizing major interests and perspectives to be included in the coalition; and (3) strategic action planning by the whole coalition.[7] These phases are essential to forming a strong coalition that has a clear sense of direction and ensures the presence of essential players. The coalition should be inclusive, with members sharing ownership and power. A successful coalition sets specific goals and develops strategies to accomplish these goals. An ongoing task of the coalition is to sustain the effort and, in effect, the coalition itself; therefore, phases two and three should be periodically revisited.

Background

The Southeastern Connecticut Indoor Air Quality Coalition is representative of a coalition with a public health focus. This coalition is a group of concerned residents, health officials, educators, business owners, parents, health-care providers, and others who are working together to improve the quality of the environment in communities, with an emphasis on indoor air quality. The coalition began in 1993 as the Community Lead Advisory Board, which addressed the issue of lead poisoning in Groton and New London, Connecticut. In January 1998, the board broadened its mission and its geographic area to become the Southeastern Connecticut Indoor Air Quality (SECT IAQ) Coalition. In August 1998, five members of the SECT IAQ participated in a national training program, sponsored by the EPA and the National Association of Counties (NACO), on developing local indoor air coalition initiatives. These members were then able to return to Connecticut and take a leadership role in further developing and enhancing the efforts of the coalition. Since the initial training, the coalition has been nationally recognized, by EPA and by NACO, for its efforts in addressing indoor air quality.

Coalition Structure

The coalition is an ad hoc group that serves southeastern Connecticut, which encompasses New London County. Membership is open to anyone who lives or works in southeastern Connecticut. Individuals and member agencies donate their time and resources and there are no membership fees. Two members serve as cochairs and two members serve as cosecretaries. Member agencies act as fiduciaries for grants. The coalition has received support and funding from NACO, the EPA, and the National Civic League. Because the SECT IAQ Coalition arose from an existing group, the coalition was able to immediately form a core initiating group that determined the focus of the coalition, conducted the stakeholder analysis, and identified and recruited other potential interested parties. The coalition used several approaches to recruit members, including brochures, advertisements on the local cable access television, and letters; however, each member of the coalition personally contacted the identified stakeholders, inviting them to attend a meeting/training on coalition building. It was at this meeting that the coalition completed vision and mission statements and began to define goals. Approximately two years after the initiating group began, and one year after the whole coalition began meeting, the coalition revisited its goals and structure. A significant change was made by replacing standing committees with temporary project-based working groups, which allowed more flexibility to members who did not have the time to devote to a committee, but wanted to be involved in promoting the coalition's mission.

Vision, Mission, Goals

The vision statement of the Southeastern Connecticut Indoor Air Quality Coalition is:

Residents of southeastern Connecticut communities will be free from illness related to indoor air pollutants. The Coalition will serve as a model for healthy communities' planning.

The mission statement of the coalition is:

To work collaboratively with the communities of southeastern Connecticut to improve the health of our residents, particularly children, through public health efforts which target indoor air quality.

Over the next several months, the coalition continued to refine its goals and develop action plans and measures. Throughout this period, the coalition conducted educational efforts for both members and the public and began laying the foundation for implementing the action plan. Programming was developed that ensured increased public awareness, such as cable television programs, developing and distributing a newsletter, and meeting with school superintendents and principals. These activities fostered a sense of accomplishment, as well as commitment, among coalition members.

From its inception in January 1998 through the spring of 2001, the coalition has made significant progress in meeting its goals. A primary goal is to improve the quality of the air in our schools, through implementation of the EPA's Tools for Schools (TFS) program throughout the region. The coalition has brought the TFS program to southeastern Connecticut and has worked with the American Lung Association (ALA) and the New London Health Department to implement a comprehensive approach to indoor air quality and asthma. This joint venture incorporates TFS, the ALA's Open Airways, and Putting on A.I.R.S. (asthma indoor risk strategies), a home assessment program developed by the New London Health Department and Lawrence and Memorial (L&M) Hospital. Coalition members have been trained in all three programs. The EPA recently recognized Edgerton Elementary School in New London for being the first to provide a comprehensive, bilingual approach to indoor air quality and asthma.

Another goal of the coalition is to reduce the number of children exposed to environmental tobacco smoke. Toward this end, the coalition developed and implemented a public awareness campaign and educational outreach effort to reduce exposure to smoke. The coalition worked closely with the local hospitals' maternity units, the Safe Children's Coalition, and local restaurants. The coalition also sponsored presentations to local schoolchildren by the "Winston Man," a former model for Winston cigarettes, on the "truth" about cigarette smoking. An important goal of the coalition is to educate the public and promote the awareness of indoor air quality and indoor environmental hazards. This is being accomplished through newsletters, information booths at health fairs and home shows, radio shows, and cable television programs. In addition, the coalition has worked with statewide groups to promote awareness of indoor air quality issues and to develop resources and programs to assist others in improving the quality of their indoor environment. Prior to the state congress going into session, the coalition held a legislative breakfast promoting increased attention to indoor air quality, which led to the introduction and support of several bills in the state legislature. This breakfast will become an annual event. Because of the rise in asthma, the coalition incorporated two goals focusing on asthma awareness and reduction: to reduce the number of people becoming ill from asthma and to work with hospitals, schools, and the community in educating people about asthma and indoor air triggers. The coalition developed and sponsored a continuing education forum for school nurses and local health care providers on childhood asthma. It also planned and implemented, with the Holleran Center for Community Action and Policy at Connecticut College, a highly successful community education day devoted to raising awareness of indoor air quality and asthma. At the legislative level, the coalition has advocated, in writing or in direct testimony, bills that promote positive changes with regard to indoor air quality or treatment of asthma.

Looking Ahead

While the SECT IAQ Coalition continues to effect positive change, it has an ongoing task of remaining vital and relevant to the needs of the community. As with any coalition, the SECT IAQ Coalition continues to strive for increased representation reflective of the community by recruiting parents, minorities, and other underrepresented groups. It is the responsibility of the coalition to continue to recruit members and replace exiting members with others who can represent their interests. For change to truly occur, it is critical to have a diverse set of needs at the table, to agree upon a common mission and common goals, and for members to have an active role in accomplishing those goals.

PEOPLE, PLACES, AND ASTHMA: ITS ECOLOGICAL IMPRINT IN OUR MIDST

Ecology can be understood as both a particular way of looking at the world and a field of study. Human ecology can be seen as an extension of the approach to the human condition. Human communities have to function in ecosystems that support and sustain them.[8] Human ecology is a multidisciplinary endeavor that looks at how an aggregate community lives within its environment, rather than as an individual phenomenon.[9] There is great diversity and inequality in the physical features and the distribution of natural resources of where populations live.[10] By taking a human ecological approach to the public health issue of asthma, one examines the how and why in communities and environments. There is increased focus regarding the prevention, etiology, epidemiology, and management of asthma. Communities where asthma rates are significant may have both limiting and supporting factors of the environment.[11] Therefore, preventing, monitoring, and treating asthma is best done through a multidisciplinary approach.

Asthma as an Increasing Public Health Concern

Asthma is defined as "a reversible airway disease that manifests as tightness in the chest with varying degrees of difficulty breathing due to either inflammation of the airways in the lungs or spasm occurring in the airways."[12]

The Centers for Disease Control (CDC) reports that 17.3 million people nationally have asthma. Connecticut hospitals supplied data for asthma-related emergency department visits and hospital admissions due to asthma. In 1998, the rate of hospital admissions across Connecticut for children 0–14 years was 22.4 per 10,000 people. The rate was 51.5 in the five largest cities. In 1998, the rate of emergency department visits in Connecticut per 10,000 people was 82.5 for children 0–4 years, 86.5 for children 5–9 years, and 71.1 for children 10–14 years.[13] Recent research on the etiology of asthma continues to point to the causes being multifactorial. These may include allergen load, commonly referred to as "asthma triggers" due to environmental exposures in ambient and indoor air, family history of asthma, diet, early respiratory infection, and atopy. Atopy is a heightened sensitivity to specific substances and is manifested by internal and external changes. The hypersensitive reaction may be distant from the region of contact with the substance responsible.[14]

The epidemiology of asthma is not exclusive to one gender, race, or class. According to the CDC, overall prevalence rates for asthma rose 75 percent during the 1980s and early 1990s. The greatest increase has been seen in children less than four years old, where the prevalence rate has increased by 168 percent. The highest incidence is in large urban centers. While rates may actually be higher in Caucasians, Hispanics and African Americans experience disproportionate death and disability due to asthma.[15] Scientists are more inclined to believe that, while race and heredity may play a part in the epidemiology of asthma, what you do and where you live may play far more significant roles.[16]

A Small City's Response to Asthma

In 1998, the city of New London ranked fifth in the state for physician-diagnosed asthma cases, with a rate of 30.3 per 10,000 people for hospital admissions and 211.4 per 10,000 people for emergency department visits. In 1999, the number of students in pre-K through 12th grade with diagnosed asthma in the New London school system was 13 percent. This figure, along with the results of the Healthy New London 2000 needs assessment, confirmed that asthma is a significant problem. In response to the high asthma rate in New London, an ad hoc group, the City Asthma Initiative, was formed to review the impact and management of asthma within the hospital and in the community. The group's members represented the New London Health Department, the Lawrence and Memorial Hospital, and the Southeastern Connecticut Visiting Nurse Association. The group's objectives were twofold: (1) to decrease utilization rates for asthma as acute care and (2) to assess and assist in mitigating environmental factors contributing to asthma in the population. This group endorsed the field testing of a pilot program, Putting on A.I.R.S., as a conceptual model for intervention and risk reduction of asthma.

Putting on A.I.R.S. is a home assessment and education program. The purpose of the Putting on A.I.R.S. program is to improve the environment in which residents with asthma live, play, attend school, and work. The program assists individuals and their families in understanding the critical importance that environmental triggers and indoor air quality play in controlling asthma. The L&M Hospital identifies families with an asthmatic member through emergency department referrals, hospital admissions departments, and school nurses. Self-referrals are accepted. Families of the patient are provided with information on the program and are offered the opportunity to participate. Following this referral, an appointment is made for a sanitarian and a public health nurse to visit the family in their home to assess the environmental triggers inside the home and environmental hazards outside the home, as necessary. An important part of the home visit may be a discussion of the home medication and treatment plan and the relationship with the patient's provider. The team provides information on abatement, provides one-on-one education about asthma triggers and prevention, and schedules a follow-up visit to the home three months after the original visit.

During the six-month pilot period of October 1, 1998, through April 5, 1999, the L&M hospital made forty-eight referrals to the program. Of these, 56 percent became active cases, 15 percent refused to take part in the program (a common problem once the crisis is over), 19 percent could not be reached (via telephone or letter), and 10 percent moved or were nonasthma cases. Several challenges to the implementation and community acceptance of this program became evident during the pilot period. These included the voluntary nature of the program, the invasive nature of a home visitation, and language and communication barriers.

The Link between Health Disparities and Environmental Injustice

A discussion of the impact of chronic diseases such as asthma illuminates the link between health disparities and environmental injustice. Can you have one without the other? Thus the question arises: Did the Putting on A.I.R.S. project address both? The objective of the Putting on A.I.R.S. project is to strengthen community prevention of asthma. In particular, it provides health care access to low-income people. In keeping with the goals of Healthy People 2010, the Putting on A.I.R.S. home intervention is a method of reducing the health disparities among special populations. The project addresses the key points cited by Healthy People 2010: "Intervention should increase access to medical care, assist with financial support for medication if necessary, determine if monitoring aids (e.g., peak flow meters) are indicated, and [determine if] environmental control measures will be essential for reducing disparities."[17]

Since 1998, the work of the Southeastern Connecticut Indoor Air Quality Coalition and the City Asthma Initiative, coupled with a new approach to viewing asthma in the community through the Putting on A.I.R.S. project, has concentrated our perspective on environmental justice. While more work is needed in surveillance efforts and raising awareness of the existence or planned placement of environmental hazards in areas where there are vulnerable populations, the initiatives conducted thus far demonstrate that programs can be tailored to help people exercise control of their environment. Viewing asthma from an ecological perspective puts the community first and may encourage a feeling of hope and investment in the environments where people live.

CONCLUSION

The link between the burden of poverty and poor health is not new to the public health community. There is increased awareness that poverty and its associated problems are among the leading risks to health. In this chapter, we have taken a systems approach to health promotion and ecology. We have attempted to show the linkages between public policies developed at the federal and state levels and their impact at the individual child and family level in the home, school, and workplace. All the programs discussed symbolize an effort to redistribute resources to eliminate health disparities. We have shown how bringing public health and economic interests together, by working with municipal government, federal and state legislators, and the community, can result in transforming Brownfields sites into healthy community spaces. We have explained how the collaborative approach used in the EPA Tools for Schools can successfully address school indoor air quality problems. We have described how community advocacy agencies, public health providers, and a city hospital work together in the Putting on A.I.R.S. program, which addresses home indoor air quality exposure problems, and have demonstrated that a modification of indoor environments can reduce the burden of asthma in the homes of New London participants.

The underlying theme across all the case studies is the importance of risk assessment in identifying problem areas, of communication and cooperation among agencies, and of collaboration through coalition building at all stages of the projects. Frequently, programs designed to address health issues are dependent on funding and are agency driven. Their weakness, and often a reason for failure of these programs in the long run, is that they do not ensure community representation at the onset of collaboration. We demonstrate how agencies, working together with the communities they serve, can interrupt the cycle of poor health through a redistribution of resources to vulnerable populations, thereby reducing health disparities among diverse social groups and promoting healthy living environments for all.

NOTES

1. Nancy Milio, *Public Health in the Market: Facing Managed Care, Lean Government, and Health Disparities* (Ann Arbor: University of Michigan, 2000), 57–59.

2. Environmental Protection Agency Brownfields Home Page, 1997, <http://www.epa.gov/swerosps/bf/glossary.htm#brow> (December 12, 2001).

3. Environmental Protection Agency Environmental Justice Home Page, 2000, <http://es.epa.gov/oeca/main/ej/faq.html> (December 12, 2001).

4. Charles Lee, "Toxics and Race," in United Church of Christ, Commission for Racial Justice, Toxic Waste and Race in the United States: A National Report on the Racial and Socio-Economic Characteristics of Communities with Hazardous Waste Sites (New York: United Church of Christ, 1987).

5. Charles Lee, *Environmental Justice, Urban Revitalization and Brownfields: The Search for Authentic Signs of Hope. A Report on the "Public Dialogues on Urban Revitalization and Brownfields: Envisioning Healthy and Sustainable Communities"* (Washington, D.C.: National Environmental Justice Advisory Council Waste and Facility Siting Subcommittee, December 1996).

6. National Civic League, *Local Indoor Air Coalition-Building Initiative Training & Resource Notebook* (Denver, Colo.: National Civic League, 1998).

7. National Civic League, *Local Indoor Air.*

8. S. Rowe, "From Reductionism to Holism in Ecology and Deep Ecology," *Ecologist* 27, no. 4 (1997): 147–51.

9. A. Hawley, "Ecology and Human Ecology," in *Origins of Human Ecology,* ed. G. C. Young (Stroudsberg, Pa.: Hutchinson Ross, 1983), 116–23.

10. C. Adams, "The Role of General Ecology to Human Ecology," in *Origins of Human Ecology,* ed. G. C. Young, 84–91 (Stroudsberg, Pa.: Hutchinson Ross, 1983).

11. A. Hawley, "Ecology and Human Ecology."

12. Department of Health and Human Services, "Healthy People 2010: Understanding and Impacting Health," DHHS Publication No. 017-001-00543-6 (Hyattsville, Md.: Department of Health and Human Services, 2000), 24–27.

13. M. Fleissner, M. Adams, and J. Kertanis, *Asthma in Connecticut* (Hartford: Connecticut State Department of Public Health, 2001), 9, 20.

14. A. Woolcock and J. Peat, "Evidence for the Increase in Asthma Worldwide," in *The Rising Trends in Asthma,* ed. Derek Chadwick and Gail Cardew, CIBA Foundation Symposium (New York: Wiley, 1997).

15. Department of Health and Human Services, "Healthy People 2010."

16. D. Schultz, *Asthma: A Public Health Partnership Tackles a Neighborhood Terror* (New York: Columbia University Press, 2001).

17. Department of Health and Human Services, "Healthy People 2010."

10

Three Political Problems for Environmental Justice

Christopher H. Foreman Jr.

Environmental justice advocacy seeks to ensure that public authorities and mainstream organized interests effectively address disproportionate burdens borne by historically disadvantaged communities.[1] Since at least the early 1980s (although its roots go back much farther) the environmental justice movement, a remarkably diverse coalition of grassroots activists and academics, has insisted that such communities are too often the "invisible man" of environmentalism. The coalition has portrayed these communities as blighted by pollution, betrayed by inadequate regulatory enforcement, and burdened by insufficient power within environmental institutions and policy processes. Native American constituencies have perceived a discrete set of assaults on both tribal sovereignty and cultural traditions, both profoundly anchored in attachment to particular lands. By the beginning of the Clinton presidency, the larger claims of inequity in siting and enforcement had been bolstered by oft-cited research findings.[2] Some suggested that disadvantaged minority communities had been actively targeted for environmental poisons, perhaps leading to higher rates of chronic illnesses, such as cancer. "Environmental racism" is the provocative label often applied to such claims. More broadly, according to one leading interpreter, environmental racism refers "to any [environmental] policy, practice, or directive that differentially affects or disadvantages . . . individuals, groups, or communities based on race or color."[3]

This chapter highlights three persistent challenges facing policy makers who seek to formulate, and activists who wish to promote, public policies that advance environmental justice. One is the difficult issue of environmental risk. Do low-income persons and persons of color bear demonstrably disproportionate shares of environmental risk? Do such persons develop pollution-related illnesses more often than other persons? And even when pollution-related health disparities cannot be demonstrated, what other kinds of environmental inequities might constitute reasonable grounds for aggressive policy remediation? A second challenge resides in the tools policymakers might bring to bear to remedy perceived injustice. How can policymakers realistically and successfully confront injustice, given the limited and constrained options at their immediate disposal? A third problem is the overall "political sustainability" of environmental justice efforts in an era prone to both unified Republican (i.e., conservative) governance and powerful environmental reform impulses that concentrate more on rationality than on democratic empowerment. What might be done to blend the rationalizing and democratizing orientations that now compete (but rarely connect effectively) in the environmental reform discourse?

HEALTH, RISK, AND ENVIRONMENTAL JUSTICE ADVOCACY

Although activists understandably have sometimes adopted a pose of certitude regarding health risks, we are far from having anything like a reliable analytic handle on such issues. Moreover, although activists have often invoked health concerns rhetorically, environmental justice has never been systematically risk driven. Probably no social movement of the sort could be, given both the limits of current knowledge and the much higher (and more accessible) political imperative to concentrate on environmental targets that offer mobilization potential.

To be sure, we are more likely to find certain environmental risks in closer proximity to poor people than to wealthier ones. A decade ago, an EPA task force on environmental equity, created in July 1990 by administrator William Reilly, determined that one problem—lead exposure—stood out in the data as a particular threat among African American youngsters.[4] There are certainly other substances having a greater cumulative adverse health impact on persons of color than on whites. Agricultural chemical exposure among Latino farmworkers is perhaps the most obvious, along with urban air toxins. The ingestion of toxic residue as a by-product of fish consumption is sometimes mentioned, but any resulting disproportionate disease incidence is not scientifically well established.[5]

An enduring challenge has been that environmental risks outside specific work settings are so often widely distributed, small, or subtle, and thus difficult to associate with discrete health effects in particular populations. Many low-income persons and persons of color reside and work (alongside a great many whites and an increasing number of suburban refugees of all colors) in cities. Anyone who lives there is almost certainly breathing dirtier air than more rural folk. Much that migrates into our urban air is troubling. Ozone, for example, is a significant respiratory irritant that can help trigger asthma attacks.[6] African Americans, young and old, suffer disproportionately from asthma, as reflected by rates of asthma-associated mortality and hospitalization.[7] But even in urban areas, both indoor and outdoor environments may be implicated in asthma incidence.[8]

At a purely impressionistic level (unimportant for scientific purposes but crucial for capturing the attention of cautious politicians and persuading them that action is imperative), it is hard to discern racial distinctiveness among the headline environmental controversies of our time. New Jersey has long been renowned for its abundance of hazardous waste, including that deposited at what was once the nation's top-ranked Superfund site at Lipari.[9] Would anyone claim that the size or configuration of that state's population of color in any way explains this? One might pose the same question about the Hanford nuclear waste site in Washington State or even the infamous Love Canal and Times Beach episodes of years ago.[10] Had Love Canal been a community of color, the past twenty years of environmental policy and environmental justice advocacy would surely have played out rather differently. The huge Fresh Kills landfill on Staten Island, the sole facility of its kind in New York City, has endured since the 1940s for many reasons (including, most recently, the need to dispose of World Trade Center wreckage), but race was not among them.[11]

The story told by the available empirical literature is more ambiguous. Many empirical studies have undercut claims of systematic ethnic bias in facility siting and in cleanup decision making, while other careful assessments verify the existence of racial and economic inequity.[12] None of this work can get far on the risk issue, suggesting that policymakers and activists alike will have to either embrace "the precautionary principle" or frame injustice in terms other than technical risk. (The "precautionary principle" holds that a product or process ought not to be used until risks associated with such use are well understood and that in the absence of such understanding we should err on the side of safety. In its strongest form, the precautionary principle would recommend against any use whatsoever until all doubt is resolved.)

Anyone who has closely watched the environmental justice movement has heard the phrase "Cancer Alley" used to describe the petrochemical industrial corridor extending from Baton Rouge to New Orleans. In movement lore, "Cancer Alley" endures as perhaps the most prominent instance of environmental harm disproportionately borne by communities of color.[13] But research has so far failed to sustain the allegation of excess cancer incidence in southern Louisiana.[14]

Actually, it is not surprising that black Louisianans have been suffering a lot from cancer, since everyone else has been as well. Science writer Michael

Fumento observes that "one fourth of us will contract cancer and one fifth of us will die of it. Indeed, as the population ages and fewer and fewer people die of other causes, more and more will die of cancer."[15] But was cancer incidence higher among black Louisianans than among other Americans? In 1990, the respective cancer incidence rates among blacks and whites nationally stood at 423 and 393 per 100,000.[16] Differences in behavior and health care access are clearly part of the explanation. Behavioral and occupational factors have been associated with cancer incidence in Louisiana, but there appears to be no overall "cancer epidemic" in that state or in "Cancer Alley."

But "Cancer Alley" endures even today in movement rhetoric, and it is not hard to understand why. A connection between petrochemical plants—or, for that matter, between any source of fearsome and unwanted material—and disease has powerful intuitive appeal for citizens, even though science may identify no causal linkage. As Howard Margolis argues, a divergence between expert and citizen perception of risk remains one of the more treacherous fault lines in environmental politics, precisely because of the profound grip that intuition wields over citizen perceptions.[17] And since one cannot prove a negative, establishing beyond all doubt that factories and dumpsites could never cause cancer, uncertainty prevails. Not surprisingly, the precautionary principle remains widely attractive.

But formal analysis, including risk assessment, is often irrelevant to or even in conflict with the political mobilization and empowerment objectives that stir activist and community enthusiasm for environmental justice. Advocates have repeatedly demanded that policy makers effectively grapple with the "multiple, cumulative and synergistic" aspects of environmental risk to low-income and minority communities. But environmental justice activists have shown little inclination to use risk assessment, however careful, to guide their advocacy priorities. Indeed, much of their rhetoric betrays the tone of hostility toward "establishment science" common in grassroots environmentalism.

Some early and oft-cited environmental justice research worked far better as political ammunition, allowing the problem of inequity to be taken seriously, than as valid and reliable analysis. Studies in the 1980s by the U.S. General Accounting Office and the United Church of Christ/Commission for Racial Justice purported to show that commercial hazardous waste facilities were more likely to be found near minority communities.[18] What is most important to grasp about these studies is not just that they were crude and potentially misleading indicators of environmental risk—though they were—but rather that they were always merely agenda-setting instruments for the movement. Analysis did not trigger activism; rather, activists needed analysis to bolster their cause. Although more refined risk analyses may have some uses in the environmental justice context, it would be naive to imagine that their conclusions will matter much to communities, unless bonded to a gratifying practical politics anchored within those communities. Careful risk analysis will be hard put to achieve this effect unless it advances political mobilization by demonstrating convincingly that minority and low-income communities are disproportionately victimized.

The environmental justice perspective became powerful not because it spoke honestly to technical questions of harm or risk—it often did not—but because it appeared to promise something larger, more uplifting, more viscerally engaging than mere careful calculation could offer. It effectively spoke to the fear and anger among local communities feeling overwhelmed by forces beyond their control and outraged by what they perceived as assaults on their collective quality of life. In this context, "multiple, cumulative, and synergistic risk" must be seen as representing a kind of technically grounded rhetoric rather than an authentic commitment to a technical perspective. Such language may seem to its users to be the price of admission to the policy process, but it most certainly is not what the ticket buyers have come to hear.

Serious environmental problems afflict low-income and minority communities. But they appear overwhelmingly to be quality-of-life problems: odors, noise, unsightly construction or destruction, traffic congestion, as well as the economic fragility that often brings people into unhappy proximity with such things. Advocacy in the "risk and racism" vein may sometimes work to trigger collateral attention to these entirely legitimate environmental issues. There are also serious health problems disproportionately affecting low-income persons and persons of color, but the potential leverage offered by pollution control over most of them ranges from slim to none. Indeed, rhetoric casting the health of low-income and minority communities in terms of pollution might conceivably be detrimental to the health of those very communities, inasmuch as it undermines attention to more significant sources of poor health.

POLICY TOOLS FOR ENVIRONMENTAL JUSTICE

Although it surely exists to some demonstrable extent, risk disparity is, in important respects, quite beside the point. Environmental justice addresses community empowerment, citizen participation, and collective representation with a broader set of social and political goals in mind. Environmental justice advocacy and policy are surely about the control of community destiny and the enhancement of personal opportunity, notwithstanding whatever risk abatement might ensue. Indeed, disadvantaged persons and communities might wish to accept some additional risk, provided that it comes attached to enhanced opportunity. Suburbs have offered increased opportunity to millions who have nevertheless had to face much that is inhospitable or inherently risky as a result of moving there, much of it related to the dominant role of the automobile in daily life. One suspects that some Native Americans would happily forgo the superior air quality of the reservation for a crack at the "dirtier" realms of Main Street or Wall Street. The most profound form of "environmental injustice" may be that few will have that choice, just as African Americans, long avoided by the white real estate market as a kind of human analogue to toxic waste, have regularly been denied a choice.

We can now look back several years to see what policies and structures have been put into place and to ask what their various payoffs have been. In many respects, the tally will leave many of us terribly dissatisfied. The major reason may be simply that the tools readily available have unavoidably been outstripped by our aspirations. Environmental justice policy has perhaps proved unsatisfying in large part because it has tackled problems that environmental and other conventional categories of policy generally are not well suited to address.

The landmark environmental legislation of the 1970s is notable for its resolute lack of focus on equity concerns. When Senator Edmund Muskie (D-Maine) pushed for tough air pollution control legislation more than thirty years ago, he conceived of air pollution as a major national problem affecting the citizenry as a whole.[19] At the time that Muskie was promoting the clean air cause, some African American leaders were notable for their lack of interest in environmental issues as they were then being discussed. The head of the National Urban League would declare that "the war on pollution is one that should be waged after the war on poverty is won."[20] This was understandable. Black leaders of the time could see no immediate reason to make air pollution protection a priority, precisely because its potential connection to the economic fortunes and life chances of their constituents appeared weak at best. Inner-city residents were the "endangered species" they cared about. Had those leaders been more attentive and supportive, they might have helped to craft legislation that would overtly have taken the interests of communities of color into account. But it is also implausible to imagine that any of the environmental legislation enacted could more effectively have targeted the fundamentally *economic* and *social* forces that ultimately underlie much present-day interest in environmental justice.

Leaders of color of that era would not have been able to bring to bear enough insight regarding how best to facilitate the systematic inclusion of, or participation by, ordinary citizens of color. Much that we know today about the capacities and limits of citizen participation in the environmental policy context we understand only because of the turbulent history of grassroots antitoxics advocacy that blossomed after Love Canal.

Exactly *how* should we foster greater environmental equity for communities of color? Activists and many policymakers can agree on this much: powerful institutions (i.e., business and government) must take seriously all community complaints and anxieties, especially regarding facility siting and environmental health. Prodded by environmental justice enthusiasts, authorities have sought myriad ways to respond: task forces, advisory panels, and the creation of an Office of Environmental Justice at EPA. All of this has attempted to give citizens some place within the governmental apparatus to lodge their grievances. The performance of such entities has been hampered by a predictable lack of authority and resources and by the difficulty of deriving a focused agenda from the myriad concerns of local groups.

Desperate for a federal statutory hook but unwilling to push for new legislation, both the EPA and the Clinton White House turned to the most plausible available prospect: Title VI of the 1964 Civil Rights Act. But progress by the end of the Clinton administration had been disappointing, a state of affairs unlikely to improve, despite good intentions and the status of Title VI as a near-sacred civil rights text. Mercifully brief, Title VI commands that "no person in the United States shall, on ground of race, color or national origin, . . . be subjected to discrimination under any program or activity receiving Federal financial assistance." It sounds simple—you can't discriminate using government money—but applied to matters environmental can lead to vast difficulties.

It is worth recalling what the creators of Title VI had in mind. In October 1947, President Truman's Committee on Civil Rights issued its historic report *To Secure These Rights*. The toughest recommendation in its thirty-three-point "program for action" was a call for withholding federal grants from educational institutions that persisted in segregationist practices.[21] Throughout the 1950s, the idea had important champions in the National Association for the Advancement of Colored People and its close congressional ally, Harlem congressman Adam Clayton Powell (D-NY). Attached to a federal school construction assistance bill, the "Powell Amendment" would be praised by some racial liberals (as necessary tough medicine) and condemned by others as a poison pill that would kill the assistance altogether. The bill did indeed fail to pass.

The idea survived to yield landmark legislative fruit in 1964. By then, Jim Crow practices were still alive but rapidly waning. One civil rights target was the Hill-Burton hospital construction grants program, first enacted in 1946. Until 1964, when its legislative authorization was amended in line with both Title VI and a federal appeals court decision, the Hill-Burton program explicitly permitted a two-track system of racially-segregated health care to flourish in southern states.[22]

When a coveted local benefit, supported by federal money (without which the benefit might not exist at all), is contaminated by overtly racist practices (e.g., racial exclusion or forced racial separation), a threat of withheld funding is a powerful and admirably direct tool. In such cases, relatively few persons today would quarrel with Title VI enforcement.

Now consider the challenge of this approach in the context of contemporary environmental policy. One must show that discrimination on the basis
of "race, color, or national origin" has occurred. Environmental justice activists have rarely been able to demonstrate in recent years the kind of
overtly racist practice that flourished with abandon during the years leading
up to the enactment of Title VI. This is unsurprising, since such practices
have drastically declined under the relentless onslaught of law, litigation, administrative policing, and public opinion. Even where authentic racial bias
continues to take a toll, as it probably does in housing and criminal justice,
the weight of official policy is usually against it.

Environmental bureaucracies have a rarely noted advantage in this respect. Relatively recent in origin, they lack long-term institutional anchors in
a racist past. The staffs of environmental agencies are doubtless markedly
more liberal than the population as a whole. Overtly racist decision making
within them would almost certainly prompt whistle blowing and investigation.

None of this is to claim that fairness inevitably prevails in environmental
decisions. Moneyed interests regularly have the same advantages in access
and expertise that they wield elsewhere in the political system. And an environmental agency cannot be above the law. Hence EPA must adhere to Title
VI and avoid dispensing federal funds in a discriminatory fashion. How,
though, should EPA accomplish this?

The agency provided a tentative answer in February 1998, when it issued
interim guidance on the application of Title VI to environmental justice. The
guidance was instantly controversial, as it paradoxically raised a host of uncertainties in the minds of many state officials and business representatives.
The Environmental Council of the States, representing the state environmental commissioners, soon opposed it as "unworkable" and lacking in a sound
scientific foundation. A revision issued for public comment in June 2000 after a considerable political battle aims for much greater specificity. Neither
community groups nor state officials have waxed enthusiastic.

There are numerous problems beyond the two fundamental realities
that environmental decisions are best construed in terms of tradeoffs
rather than "rights" and that low-income persons will always have fewer
choices than wealthier ones about where to live and what lies nearby. Activists want to protect community health, and the guidance promised that
the EPA would assess cumulative impacts on communities from new permits when Title VI was invoked in complaints to the agency. But EPA will
doubtless remain unable to slay the beast of cumulative risk for a long
time to come. Moreover, there was little or nothing explicitly about health
in the guidance.

Business-oriented conservatives suggested that one potentially ironic effect of the policy could be to further complicate the already-daunting chal-

lenge facing economic development in low-income locales. We seem to have far more anxiety than evidence on this point, but the fear is not utterly fanciful, even though conservatives have an obviously strong incentive to hype this possibility. By adding uncertainty and delay to permit approvals, the guidance could plausibly make some locales marginally less attractive to risk-averse firms. Business representatives will be quick to bring this possibility to the attention of Republican policy makers should the battle over Title VI enforcement continue.

From the perspective of an activist, there are other holes in Title VI. The EPA can influence *permits* directly but *siting* much less so. The latter is a state and local matter and the guidance was explicit about the distinction. Legal scholar James H. Colopy observes that Title VI prohibits only projects with "unjustified" disparate impacts, rather than all projects that simply have a differential impact upon one sector of a community.[23] Even casual observers of the environmental justice scene know well that what tends to motivate community protest are not "unjustified" additions to an existing pollution burden but any addition whatsoever. Yet having examined more than eighty Title VI complaints from communities around the country since 1994, EPA has found none it deems a violation. Even if the EPA eventually identifies such a violation, the ultimate remedy is withdrawal of federal funds from the offending entity. That outcome would surely prompt congressional intervention, unless the funding in question is modest, thus risking indifference by the offending party.

Activists can still reliably use Title VI to help delay siting proposals long enough to get sponsoring firms to relent. Witness the Shintech vinyl chloride plant planned for St. James Parish or a uranium enrichment facility slated for Claiborne Parish. Both Louisiana projects evaporated when exhausted sponsors pulled the plug.[24] Community groups will continue to win at this game, whether siting processes are genuinely discriminatory or not.

THE POLITICAL SUSTAINABILITY OF ENVIRONMENTAL JUSTICE

Among the few safe political predictions one could make before November 2000 was that whoever won the presidency would endorse the concept of environmental justice. For Al Gore, who became associated with the issue as a sponsor of environmental justice legislation late in his congressional career, the incentives were obvious. Moreover, in Gore, environmental justice activists had a chance for a sympathetic president who had actually thought about the subject.[25] George W. Bush, who ultimately prevailed in the presidential election, also had reason to be supportive. Anxious to please Latino and African American voters in ways that are not costly in political or fiscal terms, the Bush administration would risk appearing "racially insensitive" by overtly jettisoning environmental justice as a concern. In August 2001, EPA administrator Christine Todd Whitman (a moderate Republican who was certainly familiar with the issue as governor of New Jersey) formally made clear her embrace of environmental justice in a memorandum that could easily have been drafted by her Democratic predecessor.[26] However, the future of Title VI enforcement remains problematic if business and state regulatory authorities resist.

But a more profound challenge for environmental justice may emanate from the alternative rationalistic strain in the contemporary environmental reform discourse. It is with this subject that I began the final chapter of my book, *The Promise and Peril of Environmental Justice*. If one asks environmental justice enthusiasts to highlight one fundamental defect of environmental policy, they will likely articulate some version of the "democratic wish."[27] I speak here of the raw reality that ordinary folks are generally overmatched in the policy process by large, resource-laden entities (i.e., corporations, bureaucracies, and the large, mainstream environmental organizations).

But if one listens closely to policy research organizations and business groups active on the environmental reform front searching hard for "next generation" approaches to environmental policy, this democratizing impulse fades largely into the background. For in that world, *efficiency* is the guiding theme. Business wants less hassle, more discretion, and greater reward for a job well done. So do state-level regulators. Everyone is interested in better measures of what everyone else is achieving. Rationalizing reformers want to repair the fixation of EPA on end-of-pipe controls and "single medium" problems and to curb what Justice Stephen Breyer has called "the problem of the last ten percent."[28] By this, Breyer means abatement efforts that are unreasonably costly in light of rapidly diminishing returns. It was concern about these rationalizing arguments and the constituencies that articulated them that yielded Clinton administration efforts at innovation such as Project XL, the Common Sense Initiative, and the National Environmental Performance Partnerships. Environmentalists generally, and grassroots environmental populists in particular, were conspicuous in the indifference or outright hostility they displayed toward these efforts. Achieving an effective synthesis of the rationalizing and democratizing impulses is one of the great challenges facing environmental justice and the reform community as a whole.[29]

The policy approach and justification offered by a recent comprehensive analysis of environmental justice may be of some help.[30] Finding that significant environmental inequity does indeed burden African Americans, despite the researchers' initial doubts, they proceed to evaluate four alternative policy designs (emphasizing political mobilization, race, class, and risk) according to criteria of rationality, equity, and efficiency. They conclude that a risk-based approach scores by far the highest on the collective criteria they employ. A focus on abating the worst environmental risks almost certainly offers a significant gain, defensible in both technical and political terms, for blighted communities. Moving in this direction requires that policy makers (if not activists) go beyond making "everything" a priority, at the same time finding new ways to stimulate, support, and guide communities as they seek to identify and defend their long-term interests.

NOTES

1. Robert Bullard, ed., *Unequal Protection: Environmental Justice and Communities of Color* (San Francisco: Sierra Club Books, 1994), and Christopher H. Foreman Jr., *The Promise and Peril of Environmental Justice* (Washington, D.C.: Brookings Institution Press, 1998).

2. Marianne Lavelle and Marcia Coyle, "Unequal Protection: The Racial Divide in Environmental Law," *National Law Journal,* no. 3 (September 21, 1992): S1–S6; United Church of Christ, Commission for Racial Justice, *Toxic Wastes and Race in the United States: A National Report on Racial and Socioeconomic Characteristics of Communities with Hazardous Waste Sites* (New York: United Church of Christ, 1987).

3. Robert D. Bullard, *Dumping in Dixie: Race, Class, and Environmental Quality* (Boulder, Colo.: Westview, 1990), 98.

4. Environmental Protection Agency, *Environmental Equity: Reducing Risks for All Communities* (Washington, D.C.: June 1992).

5. Beverly H. Wright, Pat Bryant, and Robert Bullard, "Coping with Poisons in Cancer Alley," in Bullard, ed., *Unequal Protection,* 126.

6. Lester Lave, "Clean Air Sense," *Brookings Review* 15, no. 3 (Summer 1997): 40–47.

7. Centers for Disease Control and Prevention, "Children at Risk from Ozone Air Pollution—United States, 1991–1993," *Morbidity and Mortality Weekly Report* 44 (April 28, 1995): 309–312.

8. David L. Rosenstreich, et al. "The Role of Cockroach Allergy and Exposure to Cockroach Allergen in Causing Morbidity among Inner-City Children with Asthma," *New England Journal of Medicine* 336 (May 8, 1997): 1356–63.

9. Daniel Mazmanian and David Morell, *Beyond Superfailure: America's Toxics Policy for the 1990s* (Boulder, Colo.: Westview, 1992).

10. For a somewhat skeptical treatment of the Love Canal episode and other controversies, see Aaron Wildavsky, *But Is It True? A Citizen's Guide to Environmental Health and Safety Issues* (Cambridge, Mass.: Harvard University Press, 1995).

11. David Martin and Andrew C. Revkin, "As Deadline Looms for Dump, Alternative Plan Proves Elusive," *New York Times,* August, 30, 1999, A1.

12. See Foreman, *Promise and Peril,* chapter 2, for a critical assessment of early environmental justice analysis. A more recent comprehensive analysis is James P. Lester, David W. Allen, and Kelly M. Hill, *Environmental Injustice in the United States: Myths and Realities* (Boulder, Colo.: Westview, 2001).

13. See, for example, Wright et al., "Coping with Poisons."

14. Frank D. Groves et al., "Is There a 'Cancer Corridor' in Louisiana?" *Journal of the Louisiana State Medical Society* 143 (April 1996): 155–65.

15. Michael Fumento, *Science under Siege: Balancing Technology and the Environment* (New York: Morrow, 1993), 83.

16. A. Fisher, W. Worth, and D. Mayer, *Update: Is There a Cancer Epidemic in the United States?* (New York: American Council on Science and Health, 1995).

17. Howard Margolis, *Dealing with Risk: Why the Public and the Experts Disagree on Environmental Issues* (Chicago: University of Chicago Press, 1996).

18. General Accounting Office, *Siting of Hazardous Waste Landfills and Their Correlation with Racial and Economic Status of Surrounding Communities* (GAO/RCED-83-168) (Washington, D.C.: General Accounting Office, 1983). See also United Church of Christ, Commission for Racial Justice, *Toxic Wastes and Race.*

19. Alfred Marcus, "Environmental Protection Agency," in *The Politics of Regulation,* ed. James Q. Wilson, 267–303 (New York: Basic, 1980).

20. Quoted in Foreman, *Promise and Peril,* 15.

21. Charles V. Hamilton, *Adam Clayton Powell Jr.: The Political Biography of an American Dilemma* (New York: Atheneum, 1991).

22. David B. Smith, *Health Care Divided: Race and Healing a Nation* (Ann Arbor: University of Michigan Press, 1999).

23. James H. Colopy, "The Road Less Traveled: Pursuing Environmental Justice through Title VI of the Civil Rights Act of 1964," *Stanford Environmental Law Journal* 13 (1994): 125–89.

24. Foreman, *Promise and Peril,* 129.

25. See Al Gore, *Earth in the Balance: Ecology and the Human Spirit* (New York: Plume/Penguin, 1993).

26. Christine Whitman, "EPA's Commitment to Environmental Justice," memorandum, August 9, 2001, available at http://es.epa.gov/oeca/main/ej/epacommit.pdf (October 31, 2001).

27. James A. Morone, *The Democratic Wish: Popular Participation and the Limits of American Government* (New York: Basic, 1990).

28. Stephen Breyer, *Breaking the Vicious Circle: Toward Effective Risk Regulation* (Cambridge, Mass.: Harvard University Press, 1993).

29. Christopher H. Foreman Jr. "Blended Rationality and Democracy: An Elusive Synthesis for Environmental Policy Reform," *Science Communication* 20 (September 1998): 56–61.

30. Lester et al., *Environmental Justice in the United States.*

Bibliography

Adams, C. "The Role of General Ecology to Human Ecology." In *Origins of Human Ecology*, edited by G. C. Young, 84–91. Stroudsberg, Pa.: Hutchinson Ross, 1983.

Adeola, Francis O. "Environmental Hazards, Health, and Racial Inequity in Hazardous Waste Distribution." *Environment and Behavior* 26, no. 1 (January 1994): 99–126.

Albrecht, L., and R. M. Brewer. "Bridges of Power: Women's Multicultural Alliances for Social Change." In *Bridges of Power: Women's Multicultural Alliances,* edited by L. Albrecht and R. M. Brewer, 2–22. Philadelphia: New Society, 1990.

Alston, Dana. *Taking Back Our Lives: A Report to the Panos Institute on Environment, Community Development, and Race in the United States.* Washington, D.C.: Panos Institute, 1990.

———. *We Speak for Ourselves: Social Justice, Race, and Environment.* Washington, D.C.: Panos Institute, 1990.

Anderson, Andy B., Douglas L. Anderton, and John Michael Oakes. "Environmental Equity: Evaluating TSDF Siting over the Past Two Decades." *Waste Age* 25, no. 7 (July 1994): 83–100.

Anderson, Eugene N. *Ecologies of the Heart: Emotion, Belief, and the Environment.* New York: Oxford University Press, 1996.

Anderton, Douglas L. "Methodological Issues in the Spatiotemporal Analysis of Environmental Equity." *Social Science Quarterly* 77, no. 3 (1996): 508–15.

Anderton, Douglas L., Andy B. Anderson, John Michael Oakes, and Michael R. Fraser. "Environmental Equity: The Demographics of Dumping." *Demography* 31, no. 2 (May 1994): 229–48.

Anderton, Douglas L., John Michael Oakes, and Karla L. Egan. "Environmental Equity in Superfund: Demographics of the Discovery and Prioritization of Abandoned Toxic Sites." *Evaluation Review* 21, no. 1 (February 1997): 3–26.

Anderton, Douglas L., John Michael Oakes, and Michael R. Fraser. "Race, Class and the Distribution of Radioactive Wastes in Massachusetts." *New England Journal of Public Policy* 15, no. 1 (Fall/Winter 2000): 79–96.

193

Anderton, Douglas L. et al. "Hazardous Waste Facilities: 'Environmental Equity' Issues in Metropolitan Areas." *Evaluation Review* 18, no. 2 (April 1994): 123–40.

Aragón de Valdez, T. "Organizing as a Political Tool for the Chicana." *Frontiers* 5, no. 2 (1980): 7–13.

Aristotle. *Nicomachean Ethics*. Translated by Terrence Irwin. Indianapolis: Hackett, 1985.

Baksh, Michael. "Change in Machiguenga Quality of Life." In *Indigenous Peoples and the Future of Amazonia: An Ecological Anthropology of an Endangered World,* edited by Leslie E. Sponsel, 187–205. Tucson: University of Arizona Press, 1995.

Barner-Barry, Carol, and Robert Rosenwein. *Psychological Perspective on Politics*. Englewood Cliffs, N.J.: Prentice Hall, 1985.

Bass, Ronald. "Evaluating Environmental Justice under the National Environmental Policy Act." *Environment Impact Assessment Review* 18, no. 1 (1998): 83–92.

Beckerman, Stephen. "Datos Etnohistóricos Acerca de los Barí Motilones." Colección de Lenguas Indigenas, Serie Menor 7. Caracas: Montalbán, 1978.

———. "On Native American Conservation and the Tragedy of the Commons." *Current Anthropology* 37, no. 4 (1996): 659–61.

Been, Vicki. "Analyzing Evidence of Environmental Justice." *Journal of Land Use and Environmental Law* 11, no. 1 (1995): 1–36.

———. "Locally Undesirable Land Uses in Minority Neighborhoods: Disproportionate Siting or Market Dynamics?" *Yale Law Journal* 103, no. 6 (April 1994): 1383–1411.

———. "What's Fairness Got to Do with It? Environmental Justice and the Siting of Locally Undesirable Land Uses." *Cornell Law Review* 78 (1993): 1001–1085.

Been, Vicki, and Francis Gupta. "Coming to the Nuisance or Going to the Barrios? A Longitudinal Analysis of Environmental Justice Claims." *Ecology Law Quarterly* 24, no. 1 (1997): 1–56.

Behrens, Clifford A., Michael G. Baksh, and Michel Mothes. "A Regional Analysis of Barí Land Use Intensification and Its Impact on Landscape Heterogeneity." *Human Ecology* 22, no. 3 (1994): 279–316.

Black, Timothy, and John A. Stewart, "Burning and Burying in Connecticut: Are Regional Solutions to Solid Waste Disposal Equitable?" *New England Journal of Public Policy* (Spring/Summer 2001): 15–34.

Bodley, John H., ed. *Tribal Peoples and Development Issues: A Global Overview*. 3d ed. Mountain View, Calif.: Mayfield, 1988.

———. *Victims of Progress*. 3d ed. Mountain View, Calif.: Mayfield, 1990.

Boerner, Christopher, and Thomas Lambert. "Environmental Justice?" In *Center for the Study of American Business Policy Study Number* 121 (April): 16. St. Louis, Mo.: Washington University, 1994.

Bowen, William M., Mark J. Salling, Kingsley E. Haynes, and Ellen J. Cyran. "Toward Environmental Justice: Spatial Equity in Ohio and Cleveland." *Annals of the Association of American Geographers* 85, no. 4 (December 1995): 641–63.

Brandon, Katrina, Kent H. Redford, and Steven E. Sanderson, eds. *Parks in Peril: People, Politics and Protected Areas*. Washington, D.C.: Island Press and the Nature Conservancy, 1998.

Breyer, Stephen. *Breaking the Vicious Circle: Toward Effective Risk Regulation*. Cambridge, Mass.: Harvard University Press, 1993.

Brown, Lester R. "The Acceleration of History." In *State of the World,* edited by Lester R. Brown, 3–20. New York: Norton, 1996.

Brown, Phil. "Popular Epidemiology: Community Response to Toxic Waste–Induced Disease in Woburn, Massachusetts." *Science, Technology, and Human Values* (Summer/Fall 1987): 78–85.

Bryant, Bunyan. "Pollution Prevention and Participatory Research as a Methodology for Environmental Justice." *Virginia Environmental Law Journal* 14, no. 4 (1995): 589–613.

———, ed. *Environmental Justice: Issues, Policies, and Solutions.* Washington, D.C.: Island, 1995.

Bryant, Bunyan, and Paul Mohai, eds. *Race and the Incidence of Environmental Hazards: A Time for Discourse.* Boulder, Colo.: Westview, 1992.

Bullard, R. D. "Introduction." In *Confronting Environmental Racism: Voices from the Grassroots,* edited by Robert D. Bullard, 7–13. Boston, Mass.: South End, 1993.

Bullard, Robert D. "Blacks and the Environment." *Humboldt Journal of Social Relations* 14 (1987c): 165–84.

———, ed.. *Confronting Environmental Racism: Voices from the Grassroots.* Boston: South End, 1993.

———. *Dumping in Dixie: Race, Class, and Environmental Quality.* Boulder, Colo.: Westview, 1990.

———. "Endangered Environs: The Price of Unplanned Growth in Boomtown Houston." *California Sociologist* 7 (1984): 85–101.

———. "Environmental Justice for All." In *Unequal Protection: Environmental Justice and Communities of Color,* edited by Robert D. Bullard, 3–22. San Francisco: Sierra Club, 1994.

———. "Environmental Racism and 'Invisible' Communities." *West Virginia Law Review* 96 (1994): 1037–1050.

———. "Environmentalism and the Politics of Equity: Emergent Trends in the Black Community." *Phylon* 47 (1987a): 71–78.

———. "Solid Waste Sites and the Black Houston Community." *Sociological Inquiry* 53, no. 2–3 (Spring/Summer 1983): 273–88.

Bullard, Robert D., and Beverly H. Wright. "Environmental Justice for All: Community Perspectives on Health and Research Needs." *Toxicology and Industrial Health* 9, no. 5 (1993): 821–41.

———. "The Politics of Pollution: Implications for the Black Community." *Phylon* 47, no. 1 (1986): 71–78.

Burke, Laurette M. "Race and Environmental Equity: A Geographic Analysis in Los Angeles." *GEO Info System* 3, no. 9 (October 1993): 44–48.

Burn, J. M., J. W. Peltason, and T. E. Cronin. *Government by the People.* Englewood Cliffs, N.J.: Prentice Hall, 1985.

Bystydzienski, J. M. "Introduction." In *Women Transforming Politics: Worldwide Strategies for Empowerment,* edited by J. M. Bystydzienski, 1–8. Indianapolis: Indiana University Press, 1992.

Cable, S. "Women's Social Movement Involvement: The Role of Structural Availability in Recruitment and Participation Processes." *Sociological Quarterly* 33, no. 1 (1992): 35–50.

Cable, S., and C. Cable. *Environmental Problems, Grassroots Solutions: The Politics of Grassroots Environmental Conflict.* New York: St. Martin's, 1995.

Camacho, David E. *Environmental Injustices, Political Struggles: Race, Class, and the Environment.* Durham, N.C.: Duke University Press, 1998.

Capek, Stella M. "Environmental Justice Frame: A Conceptual Discussion and Application." *Social Problems* 40 (February 1993): 5–21.

Chiste, K. B. "Aboriginal Women and Self-government: Challenging Leviathan." *American Indian Culture and Research Journal* 18, no. 3 (1994): 19–43.

Churchill, W., and W. LaDuke. "The Earth Is Our Mother: Struggles for American Indian Land and Liberation in the Contemporary United States." In *The State of Native America: Genocide, Colonization, and Resistance*, edited by M. A. Jaimes, 139–88. Boston: South End, 1992.

Clark, Richard D., Steven P. Lab, and Lara Stoddard. "Environmental Equity: A Critique of the Literature." *Social Pathology* 1, no. 3 (Fall 1995): 253–69.

Clay, Jason W. "How Reserves Can Work." *Garden* 13, no. 5 (1989): 2–4.

———. *Indigenous Peoples and Tropical Forests: Models of Land Use and Management from Latin America*. Cambridge, Mass.: Cultural Survival, 1988.

———. "Indigenous Peoples: The Miner's Canary for the Twentieth Century." In *Lessons of the Rainforest*, edited by Suzanne Head and Robert Heinzman, 106–117. San Francisco: Sierra Club Books, 1990.

———. "Resource Wars: Nation and State Conflicts of the Twentieth Century." In *Who Pays the Price? The Sociocultural Context of Environmental Crisis*, edited by Barbara Rose Johnston, 19–30. Washington, D.C.: Island, 1994.

Colborn, Theo, Dianne Dumanoski, and John Peterson Myers. *Our Stolen Future*. New York: Penguin, 1997.

Colopy, James H. "The Road Less Traveled: Pursuing Environmental Justice through Title VI of the Civil Rights Act of 1964." *Stanford Environmental Law Journal* 13 (1994): 125–89.

Conroy, M., K. Kelleher, and R. Villamizar. "The Role of Population Growth in Third World Theories of Underdevelopment." In *Ethical Issues of Population Aid: Culture, Economics and International Assistance*, edited by D. Callahan and P. G. Clark, 171–206. New York: Irvington, 1981.

Contreras, R. 1993. "Toxic Survivors Seek Protection." *Voces Unidas* 3, no. 2 (1993): 7.

Contreras, R., and R. Shaw. 1993. "Isleta Pueblo's First Woman Governor." *Voces Unidas* 3, no. 2 (1993): 10 (2).

Coordinadora de las Organizaciones Indígenas de la Cuenca Amazónica. "Two Agendas on Amazon Development." *Cultural Survival Quarterly* 13, no. 4 (1989): 75–87.

Cox, Paul A. *Nafanua: Saving the Samoan Rain Forest*. New York: Freeman, 1997.

Cox, Paul Alan, and Thomas Elmqvist. "Ecocolonialism and Indigenous Knowledge Systems: Village Controlled Rainforest Preserves in Samoa." *Pacific Conservation Biology* 1 (1993): 6–13.

Cronon, William. *Changes in the Land: Indians, Colonists, and the Ecology of New England*. New York: Hill and Wang, 1983.

Cruz Takash, P. "Breaking Barriers to Representation: Chicana/Latina Elected Officials in California." *Urban Anthropology* 22, no. 3–4 (1993): 325–60.

Cushman, John Jr. "Pollution Policy Is Unfair Burden, States Tell E.P.A." *New York Times*, May 10, 1998.

Cutter, Susan L., Danika Holm, and Lloyd Clark. "The Role of Geographic Scale in Monitoring Environmental Justice." *Risk Analysis: An Official Publication of the Society for Risk Analysis* 16, no. 4 (1996): 517–26.

Daniels, N., B. Kennedy, I. Kawachi. "Why Justice Is Good for Our Health: The Social Determinants of Health Inequalities. *Daedalus* 128, no. 4 (1999): 215–52.

Dasmann, Raymond. "National Parks, Nature Conservation, and "Future Primitive.'" In *Tribal Peoples and Development Issues: A Global Overview*, edited by John H. Bodley, 301–310. Mountain View, Calif.: Mayfield, 1988.

Davidson, Pamela, and Douglas L. Anderton. "Demographics of Dumping II: A National Environmental Equity Survey and the Distribution of Hazardous Materials Handlers." *Demography* 37, no. 4 (2000): 461–66.

Di Chiro, G. "Defining Environmental Justice: Women's Voices and Grassroots Politics." *Socialist Review* 22, no. 4 (1992): 93–130.

Edelstein, Michael R. *Contaminated Communities: The Social and Psychological Impact of Residential Toxic Exposure.* Boulder: Westview, 1988.

Edwards, Audrey. "Programs That Work." *Essence* 28, no. 3 (July 1997): 42.

Ember, Lois. "IOM Report Plumbs Environmental Justice Issues." *Chemical & Engineering News* 77, no. 11 (March 1999): 14.

Fahsbender, John J. "An Analytical Approach to Defining the Affected Neighborhood in the Environmental Justice Context." *N.Y.U. Environmental Law Journal* 5 (1996): 120–80.

Feldman, M. S. *Strategies for Interpreting Qualitative Data.* London: Sage, 1995.

Fernandez, R. "Los Lunas Stops Waste Incinerator." *Voces Unidas* 3, no. 3 (1993): 12.

Fisher, A., W. Worth, and D. Mayer. *Update: Is There a Cancer Epidemic in the United States?* New York: American Council on Science and Health, 1995.

Fleissner, M., M. Adams, and J. Kertanis. *Asthma in Connecticut.* Hartford: Connecticut State Department of Public Health, 2001.

Foreman, Christopher H., Jr. "Blended Rationality and Democracy: An Elusive Synthesis for Environmental Policy Reform." *Science Communication* 20 (September 1998): 56–61.

———. *The Promise and Peril of Environmental Justice.* Washington, D.C.: Brookings Institution Press. 1998.

Fritz, Jan Marie. "Searching for Environmental Justice: National Stories, Global Possibilities." *Social Justice* 26, no. 3 (Fall 1999): 174–89.

Fugazzotto, Peter. "Angling for Environmental Justice." *Earth Island Journal* 9, no. 3 (Summer 1994): 19.

Fumento, Michael. *Science under Siege: Balancing Technology and the Environment.* New York: Morrow, 1993.

Furze, Brian, Terry De Lacy, and Jim Birchhead. *Culture, Conservation and Biodiversity: The Social Dimension of Linking Local Level Development and Conservation through Protected Areas.* New York: Wiley, 1996.

Gedicks, A. *The New Resource Wars: Native and Environmental Struggles against Multinational Corporations.* Boston: South End, 1993.

Gelobter, Michael. "Toward a Model of 'Environmental Discrimination.'" In *Race and the Incidence of Environmental Hazards,* edited by Bunyan Bryant and Paul Mohai, 64–81. Boulder, Colo.: Westview, 1992.

Geschwind, S. A. et al. "Risk of Congenital Malformations Associated with Proximity to Hazardous Waste Sites." *American Journal of Epidemiology* 135, no. 11 (1992): 1197–1207.

Bibliography

Goldman, Benjamin A. *The Truth about Where You Live: An Atlas for Action on Toxins and Morality.* New York: Times Books/Random House, 1991.

Goodman, Allen C. "A Note on Neighborhood Size and the Measurement of Segregation Indices." *Journal of Regional Science* 25 (1985): 471–76.

Gore, Al. *Earth in the Balance: Ecology and the Human Spirit.* New York: Plume/Penguin, 1993.

Gottlieb, R. *Forcing the Spring: The Transformation of the American Environmental Movement.* Washington, D.C.: Island, 1993.

Greenberg, Michael. "Proving Environmental Inequity in Siting Locally Unwanted Land Use." *Risk: Issues in Health and Safety* 14, no. 3 (Summer 1993): 235–52.

Grossman, Karl. 1992. "From Toxin Racism to Environmental Justice." *E: The Environmental Magazine* 3 (June 1992): 28–35.

———. "The People of Color Summit." In *Unequal Protection: Environmental Justice and Communities of Color,* edited by Robert Bullard. San Francisco: Sierra Club Books, 1994.

Grossman, Richard, and Gail Daneker. *Jobs and Energy.* Washington, D.C.: Environmentalists for Full Employment, 1977.

Groves, Frank, et al. "Is There a 'Cancer Corridor' in Louisiana?" *Journal of the Louisiana State Medical Society* 143 (1996): 155–65.

Guerrero, M. "Albuquerque Revisits Radioactive Dumping." *Voces Unidas* 4, no. 2 (1994): 5.

Guerrero, M., and L. Head. "Informed Collective Action—A Powerful Weapon." In *Taking Back Our Lives,* edited by D. Alston. Washington, D.C.: Panos Institute, 1990.

Hamilton, Charles V. *Adam Clayton Powell, Jr.: The Political Biography of an American Dilemma.* New York: Atheneum, 1991.

Hamilton, Laurance C. "Concern about Toxic Waste: Three Demographic Predictors." *Sociological Perspective* 28 (1985): 463–86.

Hardy-Fanta, C. *Latina Politics, Latino Politics: Gender, Culture, and Political Participation in Boston.* Philadelphia: Temple University Press, 1993.

Hawley, A. "Ecology and Human Ecology." In *Origins of Human Ecology,* edited by G. C. Young, 121–22. Stroudsberg, Pa.: Hutchinson Ross, 1983.

Head, Suzanne, and Robert Heinzman., eds. *Lessons of the Rainforest.* San Francisco: Sierra Club Books, 1990.

Hearn, Shelley A. "Tracking Toxics: Chemical Use and the Public's 'Right-to-Know.'" *Environment* 38, no. 6 (July/August 1996): 5–9, 28–30.

Hecht, Susanna. "The Sacred Cow in Green Hell." *Ecologist* 19, no. 6 (1989): 229–34.

Hecht, Susanna, and Alexander Cockburn. *The Fate of the Forest: Developers, Destroyers and Defenders of the Amazon.* New York: Harper Perennial, 1990.

Heitgerd, Janet T., Jeanne R. Burg, and Henry G. Strickland. "A Geographic Information Systems Approach to Estimating and Assessing National Priorities List Site Demographics: Racial and Hispanic Origin Composition." *International Journal of Occupational Medicine and Toxicology* 4, no. 3 (1995): 343–63.

Hershey, Marjorie R., and David B. Hill. "Is Pollution a White Thing? Racial Differences in Pre-adults Attitudes." *Public Opinion Quarterly* 41 (1977–1978): 439–58.

Higgins, Robert R. "Race and Environmental Equity: An Overview—An Environmental Justice Issue in the Policy Process." *Polity* 26, no. 2 (Winter 1993): 281–300.

Hines, Revathi I. "African Americans' Struggle for Environmental Justice and the Case of the Shintech Plant: Lessons Learned from a War Waged." *Journal of Black Studies* 31, no. 6 (July 2001): 777–89.

Hird, John A. "Environmental Policy and Equity: The Case of Superfund." *Journal of Policy Analysis and Management* 12, no. 2 (1993): 323–43.

Hird, John, and Michael Reese. "The Distribution of Environmental Quality: An Empirical Analysis." *Social Science Quarterly* 79, no. 4 (December 1998): 693–716.

Holstein, J. A., and J. F. Gubrium. *The Active Interview*. Beverly Hills: Sage, 1995.

Inter-Hemispheric Education Resource Center. "Concerned Citizens of Sunland Park." *BorderLines* 2, no. 1 (1994): 4–5.

Israel, Brian D. "An Environmental Justice Critique of Risk Assessment." *New York University Law Journal* 3, no. 2 (1994): 469–522.

Jacob, Gerald. *Site Unseen: The Politics of Siting a Nuclear Waste Repository*. Pittsburgh: University of Pittsburgh Press, 1990.

Jaimes, M. A. "American Indian Women: At the Center of Indigenous Resistance in North America." In *The State of Native America: Genocide, Colonization, and Resistance*, edited by M. A. Jaimes, 311–44. Boston: South End, 1992.

Johnston, Barbara Rose, ed. *Who Pays the Price? The Sociocultural Context of Environmental Crisis*. Washington, D.C.: Island, 1994.

Jones, M. L. "Missiles over Dineh." *Voces Unidas* 4, no. 1 (1994): 11.

Jordon, Vernono. "Energy Policy and Black People." *Vital Speeches of the Day* 45, no. 11 (March 15, 1978): 341–44.

———. "Sins of Omission." *Environmental Action* 11 (1980): 26–30.

Joseph, Mary E., and Harvey White. *A Study of the Attitudes of the Alsen Community toward the Effect of Rollins Environmental Services*. Baton Rouge, La.: School of Public Policy, Southern University, 1986.

Kaplan, T. *Crazy for Democracy: Women in Grassroots Movements*. New York: Routledge, 1997.

Kazis, Richard, and Richard Grossman. *Fear at Work: Job Blackmail, Labor, and the Environment*. New York: Pilgrim, 1983.

Kemf, Elizabeth, ed. *The Law of the Mother: Protecting Indigenous Peoples in Protected Areas*. San Francisco: Sierra Club Books, 1993.

Kozol, Jonathan. *Savage Inequalities: Children in America's Schools*. New York: Crown, 1991.

Krauss, C. "Women and Toxic Waste Protests: Race, Class and Gender as Resources of Resistance." *Qualitative Sociology* 16, no. 3 (1993): 247–62.

Krech, Shepard III. *The Ecological Indian: Myth and History*. New York: Norton, 1999.

Krieg, Eric J. "A Socio-Historical Interpretation of Toxic Waste Sites: The Case of Greater Boston." *The American Journal of Economics and Sociology* 54, no. 1 (1995): 1–14.

———. "The Two Faces of Toxic Waste: Trends in the Spread of Environmental Hazards." *Sociological Forum* 13, no. 1 (March 1998): 3–33.

Lave, Lester. "Clean Air Sense." *Brookings Review* 15 (Summer 1997): 40–47.

Lavelle, Marianne, and Marcia Coyle. "Unequal Protection: The Racial Divide in Environmental Law." *National Law Journal* 15, no. 3 (September 1992): S1–S6.

Lenssen, Nicholas. *Nuclear Waste: The Problem That Won't Go Away*. Washington, D.C.: Worldwatch Institute, 1991.

Lester, James. P., David W. Allen, and Kelly M. Hill. *Environmental Injustice in the United States: Myths and Realities.* Boulder, Colo.: Westview, 2001.

Levine, Adeline. *Love Canal: Science, Politics, and People.* Lexington, Mass.: Lexington, 1982.

Lizarralde, Manuel. "The Barí Responses to Mineral and Lumber Concessions in Venezuela." In *El Dorado Revisited: Gold, Oil, Environment, People and Rights in the Amazon,* edited by Leslie Sponsel. Washington, D.C.: Island, forthcoming.

———. "Biodiversity and the Loss of Indigenous Languages and Knowledge in South America." In *On Biocultural Diversity: Linking Languages, Knowledge and the Environments,* edited by Luisa Maffi, 265–81. Washington: Smithsonian Institution Press, 2001.

———. "Ethnoecology of Monkeys among the Barí of Venezuela: Perception, Use and Conservation." In *Primates Face to Face: The Conservation Implications of Human and Nonhuman Primate Interactions,* edited by Agustin Fuentes and Linda D. Wolfe, 85–100. Cambridge: Cambridge University Press, 2002.

———. "500 Years of Invasion: Eco-Colonialism in Indigenous Venezuela." *Kroeber Anthropological Society Papers* 75–76 (1992): 62–79.

———. "Indigenous Knowledge and Conservation of the Rainforest: Ethnobotany of the Barí of Venezuela." In *Ethnobotany and Conservation of Biocultural Diversity,* edited by Thomas Carlson and Luisa Maffi. New York: New York Botanical Garden, forthcoming.

Lizarralde, Roberto. "Barí Settlement Patterns." *Human Ecology* 19, no. 4 (1991): 428–52.

Lizarralde, Roberto, and Stephen Beckerman. "Historia Contemporánea de los Barí." *Antropológica* 58 (1982): 3–51.

Lofland, J., and L. Lofland. *Analyzing Social Settings: A Guide to Qualitative Observation and Analysis.* 3d ed. Belmont, Calif.: Wadsworth, 1995.

Lopez Zent, Egleé, and Stanford Zent. "Amazonian Indians as Ecological Disturbance Agents: The Hoti of Venezuela." In *Ethnobotany and Conservation of Biocultural Diversity,* edited by Thomas Carlson and Luisa Maffi. New York: New York Botanical Garden, forthcoming.

Lord, Charles P., and William A. Shutkin. "Environmental Justice and the Use of History." *Environmental Affairs* 22 (1994): 1–26.

Maffi, Luisa, ed. *On Biocultural Diversity: Linking Languages, Knowledge and the Environments.* Washington, D.C.: Smithsonian Institution Press, 2001.

Manaster, K. A. *Environmental Protection and Justice.* Cincinnati: Anderson, 2000.

Marcus, Alfred. "Environmental Protection Agency." In *The Politics of Regulation,* edited by James Q. Wilson. New York: Basic, 1980.

Margolis, Howard. *Dealing with Risk: Why the Public and the Experts Disagree on Environmental Issues.* Chicago: University of Chicago Press, 1996.

Markham, William, and Eric Ruffa. "Class, Race, and the Disposal of Urban Waste: Locations of Landfills, Incinerators, and Sewage Treatment Plants." *Sociological Spectrum* 17, no. 2 (April–June 1997): 235–48.

Marquez, Benjamin. "Mobilizing for Environmental and Economic Justice: The Mexican-American Environmental Justice Movement." *Capitalism, Nature, Socialism* 9, no. 4 (December 1998): 36, 43–60.

Martinez, Dennis. "First People, Firsthand Knowledge." *Sierra* 81, no. 6 (1996): 50–51, 70–71.

Massey, Douglas S., and Nancy A. Denton. "Residential Segregation of Blacks, His-panics, and Asians by Socioeconomic Status and Generation." *Social Science Quarterly* 69, no. 4 (December 1988): 797–817.

Mazmanian, Daniel, and David Morell. *Beyond Superfailure: America's Toxics Policy for the 1990s.* Boulder: Colo.: Westview, 1992.

———. "The 'NIMBY' Syndrome: Facility Siting and the Failure of Democratic Dis-course." In *Environmental Policy in the 1990s,* edited by Norman J. Vig and Michael E. Kraft, 125–44. Washington, D.C.: CQ Press, 1990.

McCaull, Julian. "Discriminatory Air Pollution: If the Poor Don't Breathe." *Environ-ment* 18 (1976): 26–32.

McCay, Bonnie J., and James M. Acheson, eds. *The Question of the Commons: The Cul-ture and Ecology of Communal Resources.* Tucson: University of Arizona Press, 1987.

McCoy, M. "Gender or Ethnicity: What Makes a Difference? A Study of Women Tribal Leaders." *Women & Politics* 12, no. 3 (1992): 57–68.

McCraken, G. *The Long Interview.* Newbury Park, Calif.: Sage, 1988.

McGlen, N. E., and K. O'Connor. *Women, Politics, and American Society.* Englewood Cliffs, N.J.: Prentice-Hall, 1995.

Miles, M. B., and A. M. Huberman. *Qualitative Data Analysis,* 2d ed. London: Sage, 1994.

Milio, Nancy. *Public Health in the Market: Facing Managed Care, Lean Government, and Health Disparities.* Ann Arbor: University of Michigan Press, 2000.

Milius, Susan. "When Worlds Collide: Why Can't Conservation Scientists and Indige-nous Peoples Just Get Along?" *Science News* 154, no. 6 (1998): 92–94.

Milton, Katharine. "Civilization and Its Discontents: Amazonian Indians Experience the Thin Wedge of Materialism." *Natural History* 3 (1992): 36–45.

Moberg, Mark. "Co-Opting Justice: Transformation of a Multiracial Environmental Coali-tion in Southern Alabama." *Human Organization* 60, no. 2 (Summer 2001): 166–77.

Mohai, Paul. "The Demographics of Dumping Revisited: Examining the Impact of Al-ternate Methodologies in Environmental Justice Research." *Virginia Environmen-tal Law Journal* 14 (1995): 615–53.

Mohai, Paul, and Bunyan Bryant. "Environmental Racism: Reviewing the Evidence." In *Race and the Incidence of Environmental Hazards: A Time for Discourse,* ed-ited by Bunyan Bryant and Paul Mohai, 163–76. Boulder, Colo.: Westview, 1992.

Mohai, Paul, and David Kershner. "Race and Environmental Voting in the U.S. Con-gress." *Social Science Quarterly,* 2002.

Morgen, S., and A. Bookman. "Rethinking Women and Politics: An Introductory Es-say." In *Women and the Politics of Empowerment,* edited by A. Bookman and S. Morgen, 3–29. Philadelphia: Temple University Press, 1988.

Morone, James A. *The Democratic Wish: Popular Participation and the Limits of American Government.* New York: Basic, 1990.

Myers, Norman. *The Primary Source: Tropical Forests and Our Future.* New York: Norton, 1992.

Naples, N. A. "Activist Mothering: Cross-Generational Continuity in the Community Work of Women from Low-Income Urban Neighborhoods." *Gender and Society* 6, no. 3 (1992): 441–63.

———. "Women's Community Activism: Exploring the Dynamics of Politicization and Diversity." In *Community Activism and Feminist Politics: Organizing across Race, Class, and Gender,* edited by N. A. Naples, 327–49. New York: Routledge, 1998.

Naples, N. A., and L. Naranjo. "Martineztown Defeats Courthouse!" *Voces Unidas* 4, no. 2 (1994): 4.

Nietschmann, Bernard. *Between Land and Water: The Subsistence Ecology of the Miskito Indians, Eastern Nicaragua.* New York: Seminar, 1973.

———. "Hunting and Fishing Focus among the Miskito Indians, Eastern Nicaragua." *Human Ecology* 1 (1972): 41–67.

Novotny, Patrick. "Where We Live, Work and Play: Reframing the Cultural Landscape of Environmentalism in the Environmental Justice Movement." *New Political Science* 32 (Summer 1995): 61–79.

Nozick, R. *Anarchy, State and Utopia.* New York: Basic, 1974.

Oakes, John Michael, Douglas L. Anderton, and Andy B. Anderson. "A Longitudinal Analysis of Environmental Equity in Communities with Hazardous Waste Facilities." *Social Science Research* 25 (1996): 125–48.

Oldfield, M. L., and J. B. Alcorn, eds. *Biodiversity: Culture, Conservation and Ecodevelopment.* Boulder, Colo.: Westview, 1991.

Orleck, A. "Tradition Unbound: Radical Mothers in International Perspective." In *The Politics of Motherhood: Activist Voices from Left to Right,* edited by A. Jetter, A. Orleck, and D. Taylor, 3–11. Hanover, N.H.: University Press of New England, 1997.

Oviedo, Gonzalo. *Los Pueblos Indígenas y la Conservación: Declaración de Principios del WWF.* Gland, Switzerland: World Wildlife Fund, 1996.

Paehlke, R., and P. Vaillancourt Rosneau. "Environment/Equity: Tensions in North American Politics." *Policy Studies Journal* 21, no. 4 (1993): 672–86.

Pardo, M. "Doing It for the Kids: Mexican American Activists, Border Feminists?" In *Feminist Organizations: Harvest of the Women's Movement,* edited by M. M. Ferree and P. Y. Martin, 356–71. Philadelphia: Temple University Press, 1995.

Partners in the Environment. "New Mexico: The Land of Enchantment?" *EAGLE* 1, no. 6 (1993): 5 (2).

Pellizzari, E. F., R. I. Perritt, and C. A. Clayton. "National Human Exposure Assessment Survey (NHEXAS): Exploratory Survey of Exposure among Population Subgroups in EPA Region V." *Journal of Exposure Analysis and Environmental Epidemiology* 9, no. 49 (1999): 55.

Peluso, Nancy Lee. *Rich Forest, Poor People: Resource Control Resistance in Java.* Berkeley: University of California Press, 1992.

Peña, D. "The 'Brown' and the 'Green:' Chicanos and Environmental Politics in the Upper Rio Grande." *Capitalism Nature Socialism: A Journal of Socialist Ecology* 3, no. 1 (1992): 79–103.

Phillips, Carl V., and Ken Sexton. "Science and Policy Implications of Defining Environmental Justice." *Journal of Exposure Analysis and Environmental Epidemiology* 9 (1999): 9–17.

Pollock, Phillip H. III, and M. Elliot Vittas. "Who Bears the Burdens of Environmental Pollution? Race, Ethnicity and Environmental Equity in Florida." *Social Science Quarterly* 76, no. 2 (June 1995): 294–310.

Ponting, Clive. *A Green History of the World: The Environment and the Collapse of Great Civilizations.* New York: Penguin, 1991.

Posey, Darrell A. *Biological and Cultural Diversity: The Inextricable, Linked by Language and Politics.* Oxford, England: Oxford Centre for the Environment, Ethics and Society at Mansfield College, 1996.

———. "Interpreting and Applying the 'Reality' of Indigenous Concepts: What Is Necessary to Learn From the Natives?" In *Conservation in Neotropical Forests: Working from Traditional Resources Use,* edited by Kent H. Redford and Christine Padoch, 21–34. New York: Columbia University Press, 1992.

Prindeville, D. M., and J. G. Bretting. "Indigenous Women Activists and Political Participation: The Case of Environmental Justice." *Women & Politics* 19, no. 1 (1998): 39–58.

Pulido, L. "Sustainable Development at Ganados del Valle." In *Confronting Environmental Racism: Voices from the Grassroots,* edited by R. D. Bullard, 123–39. Boston: South End, 1993.

Ramphal, Shridath, and Steven W. Sinding. "Conclusions." In *Population Growth and Environmental Issues,* edited by Shridath Ramphal and Steven W. Sinding. Westport, Conn.: Praeger, 1996.

Ratcliffe, J. "Poverty, Politics and Fertility: The Anomaly of Kerala." *Hastings Center Report* 7, no.1 (1977): 34–42.

Rawls, J. *A Theory of Justice.* Cambridge, Mass.: Harvard University Press, 1971.

Reaka-Kudla, Marjorie L., Don E. Wilson, and Edward O. Wilson, eds. *Biodiversity II: Understanding and Protecting Our Biological Resources.* Washington, D.C.: Joseph Henry, 1997.

Redford, Kent H. "The Ecologically Noble Savage." *Cultural Survival Quarterly* 15, no. 1 (1990): 46–48.

Redford, Kent H., and Christine Padoch, eds. *Conservation in Neotropical Forests: Working from Traditional Resources Use.* New York: Columbia University Press, 1992.

Redford, Kent H., and Allyn M. Stearman. "Forest-Dwelling Native Amazonians and the Conservation of Biodiversity: Interests in Common or in Collision?" *Conservation Biology* 7, no. 2 (1993): 248–55.

Reed, Richard K. *Forest Dwellers, Forest Protectors: Indigenous Models for International Development.* Boston: Allyn and Bacon, 1997.

Richards, Paul W. *The Tropical Rain Forest: An Ecological Study.* 2d ed. Cambridge: Cambridge University Press, 1996.

Ringquist, Evan. "Equity and the Distribution of Environmental Risk: The Case of the TRI Facilities." *Social Science Quarterly* 78, no. 4 (December 1997): 811–29.

Ripley, Randall, and Grace A. Franklin. *Congress, the Bureaucracy, and Public Policy.* Homewood, Ill: Dorsey, 1980.

Rodriguez, S. "Land, Water, and Ethnic Identity in Taos." In *Land, Water, and Culture: New Perspectives on Hispanic Land Grants,* edited by C. L. Briggs and J. R. Van Ness, 313–403. Albuquerque: University of New Mexico Press, 1987.

Romero, F. "Grupitos: Citizen Involvement at the Kitchen Table." *Nuestro Pueblo* 7, no. 2 (Summer 1997).

Rosenbaum, Walter A. *Environmental Politics and Policy.* Washington, D.C.: CQ Press, 1991.

Rosenstreich, David, et al. "The Role of Cockroach Allergy and Exposure to Cockroach Allergen in Causing Morbidity among Inner-City Children with Asthma," *New England Journal of Medicine* 336 (May 8, 1997): 1356–63.

Ross, Heather E. "Using NEPA in the Fight for Environmental Justice." *William and Mary Journal of Environmental Law* 18 (1994): 353–74.

Rowe, S. "From Reductionism to Holism in Ecology and Deep Ecology," *Ecologist* 27, no. 4 (1997): 147–51.

Russell, Dick. "Environmental Racism: Minority Communities and Their Battles Against Toxics." *Amicus Journal* 11, no. 2 (Spring 1989): 22–32.

Sahlins, Marshall. "Notes on the Original Affluent Society." In *Man the Hunter,* edited by Richard Lee and Irven DeVore, 85–89. New York: Aldine, 1968.

Schlosberg, David. "Networks and Mobile Arrangements: Organizational Innovation in the US Environmental Justice Movement." *Environmental Politics* 8, no. 1 (Spring 1999): 122–48.

Schmink, Marianne, Kent H. Redford, and Christine Padoch. "Traditional Peoples and the Biosphere: Framing the Issues and Defining the Terms." In *Conservation in Neotropical Forests: Working from Traditional Resources Use,* edited by Kent H. Redford and Christine Padoch, 3–34. New York: Columbia University Press, 1992.

Schultz, D. *Asthma: A Public Health Partnership Tackles a Neighborhood Terror.* New York: Columbia University Press, 2001.

Scott, Michael S., and Susan L. Cutter. "Using Relative Risk Indicators to Disclose Toxic Hazard Information to Communities." *Cartography and Geographic Information Systems* 24, no. 3 (1997): 158–71.

Sen, A. *Development as Freedom.* New York: Anchor, 1999.

Sexton, Ken, and John L. Adgate. "Looking at Environmental Justice from an Environmental Health Perspective." *Journal of Exposure Analysis and Environmental Epidemiology* 9 (1999): 3–8.

Sheppard, Eric, Helga Leitner, Robert B. McMaster, and Hongguo Tian. "GIS-Based Measures of Environmental Equity: Exploring Their Sensitivity and Significance." *Journal of Exposure Analysis and Environmental Epidemiology* 9 (1999): 18–28.

Shiva, Vandana et al. *Biodiversity: Social and Ecological Perspectives.* London: Zed, 1991.

Smith, David B. *Health Care Divided: Race and Healing a Nation.* Ann Arbor, Michigan: University of Michigan Press, 1999.

Smith, Rhonda. "MOSES Leads Winona, Texas, to Environmental Justice." *Crisis (The New)* 104, no. 1 (July 1997): 30–31.

South West Organizing Project. "Carletta Tilousi: Fighting for Her People." *Voces Unidas* 1, no. 4 (1991): 13.

———. "Five Years Later: SWOP's Letter to the 'Group of Ten' Revisited." *Voces Unidas* 5, no. 3 (1995), special insert: 1–4.

———. "Grassroots Democracy in Action: An Interview with Richard Moore." *Voces Unidas* 4, no. 3 (1994): 9 (3).

———. "The Selling of New Mexico." *Voces Unidas* 4, no. 2 (1994a), special insert.

Sponsel, Leslie E., ed. *Indigenous Peoples and the Future of Amazonia: An Ecological Anthropology of an Endangered World.* Tucson: University of Arizona Press, 1995.

Sponsel, Leslie E., Robert C. Bailey, and Thomas N. Headland. "Anthropological Perspectives on the Causes, Consequences, and Solutions of Deforestation." In *Tropical Deforestation: The Human Dimension,* edited by Leslie E. Sponsel, Thomas N. Headland, and Robert C. Bailey, 3–52. New York: Columbia University Press, 1996.

Steingraber, S. *Living Downstream.* New York: Vintage, 1997.

Stevens, Stan, ed. *Conservation through Cultural Survival: Indigenous Peoples and Protected Areas.* Washington, D.C.: Island, 1997.

Strauss, A., and J. Corbin. *Basics of Qualitative Research: Grounded Theory Procedures and Techniques.* London: Sage, 1990.

Stretesky, Paul, and Michael J. Hogan. "Environmental Justice: An Analysis of Superfund Sites in Florida." *Social Problems* 45, no.2 (May 1998): 268–87.

Taliman, V. "Uranium Miners Made Profits, Left Town, and Stuck Natives with Open Pits and Contamination Worries." *Voces Unidas* 1, no. 3 (1991): 6.

Taylor, D. "Environmentalism and the Politics of Inclusion." In *Confronting Environmental Racism: Voices from the Grassroots,* edited by R. D. Bullard, 53–61. Boston: South End, 1993.

Taylor, L., et al. "The Importance of Cross-Cultural Communication between Environmentalists and Land-Based People." *Workbook* 13, no. 3 (1998): 90–93.

Terborgh, John. *Requiem for Nature.* Washington, D.C.: Island, 1999.

Terry, Larry. "Activism on the Bayou." *Chemical Week* 161, no. 20 (May 1999): 59.

Thomas, S. *How Women Legislate.* New York: Oxford University Press, 1994.

Tilly, L. A., and P. Gurin. "Women, Politics and Change." In *Women, Politics and Change,* edited by L. A. Tilly and P. Gurin, 3–34. New York: Russell Sage Foundation, 1990.

Toledo, Victor Manuel. "What Is Ethnoecology? Origins, Scope, and Implications of a Rising Discipline." *Etnoecología* 1, no. 1 (1992): 5–21.

United Church of Christ, Commission on Racial Justice. *Toxic Wastes and Race in the United States: A National Report on the Racial and Socioeconomic Characteristics of Communities with Hazardous Waste Sites.* New York: United Church of Christ, 1987.

United Nations Development Program, "Human Development Report." New York: Oxford University Press, 1990.

Urban Environment Conference, Inc. *Taking Back Our Health: An Institute on Surviving the Threat to Minority Communities.* Washington, D.C.: Urban Environment, Inc., 1985.

U.S. Department of Commerce, Bureau of Census. "1990 Census of Population— General Population Characteristics, New Mexico, 1990. CP–1–33.

U.S. Department of Health and Human Services. "Healthy People 2010: Understanding and Impacting Health," Hyattsville, Md.: DHHS Publication No. 017–001–00543–6, 2000.

U.S. Department of Health and Human Services, Centers for Disease Control. "Children at Risk from Ozone Air Pollution—United States, 1991–1993." *Morbidity and Mortality Weekly Report* 44, no. 28 (April 1995): 309–12.

U.S. Environmental Protection Agency. "Environmental Equality: Reducing Risk for All Communities." 2 vols. EPA230-R-92-008 and EPA230-R-92-008A. Washington, D.C.: Policy, Planning and Evaluation, Environmental Protection Agency, 1992.

U.S. General Accounting Office. "Hazardous and Non-hazardous Waste: Demographics of People Living Near Waste Facilities." Washington D.C.: General Accounting Office, 1995.

U.S. General Accounting Office. "Siting of Hazardous Waste Landfills and Their Correlation with Racial and Economic Status of Surrounding Communities." Washington, D.C.: General Accounting Office, 1983.

Ventocilla, Jorge, Heraclio Herrera, and Valerio Núñez. *Plants and Animals in the Life of the Kuna.* Austin: University of Texas Press, 1995.

Vickers, William T. "Hunting Yields and Game Composition over Ten Years in an Amazon Indian Territory." In *Neotropical Wildlife Use and Conservation,* edited by J. G. Robinson and K. H. Redford, 53–81. Chicago: University of Chicago Press, 1991.

Waldmann, R. J. "Income Distribution and Infant Mortality." *Journal of Economics* 107 (1992): 1283–1302.

Waller, Lance A., Thomas A. Louis, and Bradley P. Carlin. "Environmental Justice and Statistical Summaries of Differences in Exposure Distributions." *Journal of Exposure Analysis and Environmental Epidemiology* 9 (1999): 56–65.

West, Patrick C., et al. "Minority Anglers and Toxic Fish Consumption: Evidence of the State-Wide Survey of Michigan." In *Race and the Incidence of Environmental Hazards: A Time for Discourse,* edited by Bunyan Bryant and Paul Mohai, 100–113. Boulder, Colo.: Westview, 1992.

White, Harvey L. "Environmental Justice as a Public Health Imperative." In *Health Praxis and Minority Communities,* edited by Harvey L. White and Angela F. Ford. Pittsburgh: University of Pittsburgh Center for Minority Health, 2001.

———. "Hazardous Waste Incineration and Minority Communities." In *Race and the Incidence of Environmental Hazards,* edited by Bunyan Bryant and Paul Mohai, 126–39. Boulder, Colo.: Westview, 1992.

———. "Political Struggles: Race, Class and the Environment." *Environmental Injustices.* Durham, N.C.: Duke University Press, 1998.

Wildavsky, Aaron. *But Is It True? A Citizen's Guide to Environmental Health and Safety Issues.* Cambridge Mass.: Harvard University Press, 1995.

Wilkinson, R. *Unhealthy Societies: The Afflictions of Inequality.* New York: Routledge, 1996.

Wilson, Edward O. *The Diversity of Life.* New York: Norton, 1992.

Woolcock, A., and J. Peat. "Evidence for the Increase in Asthma Worldwide." In *The Rising Trends in Asthma,* edited by Derek Chadwick and Gail Cardew. CIBA Foundation Symposium, 206. New York: Wiley, 1997.

Wright, Beverly H., Pat Bryant, and Robert D. Bullard. "Coping with Poisons in Cancer Alley." In *Unequal Protection: Environmental Justice and Communities of Color,* edited by Robert D. Bullard. San Francisco: Sierra Club Books, 1994.

Wright, R. *Native Son.* New York: Harper & Brothers, 1940.

Yandle, Tracy, and Dudley Burton. "Reexamining Environmental Justice: A Statistical Analysis of Historical Waste Landfill Siting Patterns in Metropolitan Texas." *Social Science Quarterly* 77, no. 3 (September 1996): 477–92.

Zimmerman, Rae. "Social Equity and Environmental Risk." *Risk Analysis* 13, no. 6 (December 1993): 649–66.

Index

212 *Index*

House Subcommittee on Transportation
and Hazardous Materials, 14
housing, 12–13, 64, 69–70, 74–77
Houston, Tex., 64–65, 87
H.R. 1924, 15
Human Development Report (United
Nations), 31–32
Hunt, Jim, 7
hypotheses, ambiguous, 84–90, 98

incineration, hazardous waste, 120–22
income, 12–13, 76, 91. *See also* low-
income people; wealth
Indigenous Environmental Network,
22n26
indigenous people, 39–44, 50–51,
54n17. *See also* Barí people
indoor air quality, 172–78
"Indoor Air Quality for Connecticut's
Schools: Toward Healthy and Safe
School Environments," 170
indoor air quality (IAQ), 166–74
Indoor Air Quality Tools for Schools
Action Kit, 166
"Industrial and Hazardous Waste
Management Firms" (Environmental
Services Ltd.), 103n55
industry: assessments impact on
environment, 10; community
attitudes toward, 15–16, 89–90; on
environmental justice movement,
9–11; locations of, 22n29; and
research on risk and siting, 95, 98
inequality, 26–30, 36. *See also*
environmental inequality
injustice, 28–29
Institute of National Park (IMPARQUES),
50
Interim National Black Environmental
and Economic Justice Coordinating
Committee, 18
intermediate processing centers, 67–68,
72–73
international environmentalism, 61
Isleta, Pueblo, 140

Jackson, Jesse, 9, 21n20
Jacob, Gerald, 117

Japan, 10
Jennings, Cynthia, 126, 128–30
Jesuit missionaries, 47
jet-engine power generator, 130, 131
Johannesburg, 19
Jordan, Vernon, 9

Kennedy Commission, 96–97
Kerala, India, 30–31
King, Martin Luther Jr., 5, 21n20
Kirtland Air Force Base, 140
knowledge and information gap, 119,
122
Komer, Odessa, 6
Krech, Shepard, III, 41–42
Krieg, Eric J., 86
Kuna, 55n28, 57n50
Kuna Yala, 55n28

LaDuke, Winona, 151
landfills: in Connecticut, xv, 67–68, 71,
77, 128–29; in New York, 20n17, 183;
nonhazardous, 66; in southern states,
ix–x; United Church of Christ on, x,
8, 184
Lasswell, Harold, 111
Latinos. *See* Hispanics
Lawrence and Memorial Hospital, 174,
176–77
lead poisoning, 14, 135, 172, 182
League of Conservative Voters, 9
Lee, Charles, 11, 19n2, 163
legislation, 14–15, 186–91
Lenssen, Nicholas, 117
Lewis, John, 14–15, 19n2
libertarian political philosophy, 28–29
licensed environmental professionals
(LEPs), 163
life expectancy, 32
Limits to Growth (Meadows), 5
Lipari, 183
Lizarralde, Roberto, 55n29
locally undesirable land uses (LULUs),
x, 12–13, 27, 65
logging, 48
longitudinal endogeneity, 89–90, 98
longitudinal research design, 88–90, 98

61–66, 68–69, 72–78, 90–93, 97–98;
types in Connecticut, 67
Waste Isolation Pilot Project (WIPP), 139
waste streams, assessment of, 66
water area, 70, 74–75, 77
Waterbury, Conn., 131, *133*
water quality, 152
Wayne County, Mich., 16–17
Wayne State University, 16
wealth, 31–32, 64. *See also* low-income
people
West Indian mahogany, 48
White, Leon, 21n21
Whitman, Christine Todd, 190
Why In My Back Yard? (WIMBY)
syndrome, xiv, 112, 119–23
Wilkinson, Richard, 36
"Winston Man," 174
Woburn, Mass., 92
women, 139–54; in environmental
justice movement, 140; ideologies
of, 147–51; paths to leadership,
145–46; policy initiatives of, 151–53;

political activism of, xv, 141–42,
146–47
Woodcock, Leonard, 6
"Working for Environmental and
Economic Justice and Jobs"
(conference), 6, 10
World Population Conference (1974), 30
World Summit on Sustainable
Development, 19
World Trade Center, 183
Wright, Beverly, 19n2
Wright, Richard, 34

Xavier University (New Orleans), 20n13

Yale University, 167, 168, 171
Young, Whitney, 5
Younger, James, 126–28
Yucca Mountain, Nev., 115, 117

Zent, Stanford, 43
zip codes, 62–64, 68, 85–87, 93–94
zoning decisions, 89–90, 98

About the Contributors

Timothy Black

Associate professor of sociology, director of the Center for Social Research, University of Hartford

Dr. Black is an associate professor of sociology and the director of the Center for Social Research at the University of Hartford. His areas of research include urban poverty, social welfare, and human service delivery systems. He is currently working on a book about economically and socially marginalized Puerto Rican men in Springfield, Massachusetts.

Estelle Bogdonoff

Cochair, Southeastern Connecticut Indoor Air Quality Coalition

Estelle Bogdonoff is a certified health education specialist with extensive experience in community health education program development, evaluation, and administration. Recognized for her local public health leadership and creativity, Bogdonoff has been active in building collaborations and coalitions that address community health issues. She currently provides consulting services related to coalition building, community health education programs and evaluation, and training.

Bunyan Bryant

Director, Environmental Justice Initiative, School of Natural Resources and Environment, University of Michigan

Dr. Bryant is a member of the League of Conservation Voters and the U.S. Environmental Protection Agency's Clean Air Act Advisory Committee. Currently, he teaches two environmental justice courses and speaks at college campuses and professional conferences throughout the nation. He is at the forefront of an environmental justice initiative within the school. His latest interest is global climate change and environmental justice. Bunyan Bryant believes that no other issue threatens developing countries and low-income and people-of-color communities in developed nations more than global climate change.

Kathleen Cooper-McDermott
Ledge Light Health District, Groton
A public health nurse, Kathleen Cooper-McDermott's professional and social path has led her to become aware of the significance of fostering and creating healthy communities. Her focus and contribution is in the area of the ecology of health and wellness of populations and the social ecology of public health nursing to promote this. She has her master of science in nursing and master of public health degrees from the University of Connecticut.

Pamela R. Davidson
University of Wisconsin, Madison
Dr. Davidson is currently a postdoctoral research fellow at the Center for Demography and Ecology at the University of Wisconsin. Her research focuses on sociological factors associated with environmental justice outcomes. Future research plans include an examination of the health implications and ecological context of hazardous waste and industrial facility siting patterns.

Christopher H. Foreman Jr.
Professor of public policy, School of Public Affairs, University of Maryland at College Park
Until recently, Dr. Foreman was a senior fellow in the Governmental Studies program at the Brookings Institution. His writings have focused mainly on the politics of regulation, public health, and governmental reform. His active research interests include the politics of tobacco, the politics of environmental reform, policy making for minority health, and the challenge of strategic workforce planning in health and safety programs.

Kenny Foscue
Connecticut Department of Public Health, Division of Environmental Epidemiology and Occupational Health
Kenny Foscue coordinates health education, risk communication, and community involvement programs for the division as part of the Cooperative Agreement program with the federal Agency for Toxic Substances and Disease Registry. He has a master of public health degree from the University of Connecticut and a bachelor of arts in sociology from the University of North Carolina.

Cynthia R. Jennings
Board chairperson, Connecticut Coalition for Environmental Justice
A graduate of the University of Connecticut law school, Cynthia Jennings currently practices environmental law. Her law office, which is located in Bridgeport, Connecticut, takes on human rights and civil liberties cases.

Manuel Lizarralde
Assistant professor anthropology/botany, Connecticut College
A native of Venezuela, Dr. Lizarralde has focused much of his work on the relationship of indigenous Latin Americans to the environment, including the types of areas indigenous peoples inhabit and their use of plants. He draws from his background in botany, geography, and anthropology in his classes on indigenous people and their use of tropical rain forests. He teaches classes on ethnobotany and ecological anthropology.

Mark Mitchell
Director, Connecticut Coalition for Environmental Justice, Hartford
Dr. Mitchell is a physician specializing in epidemiology and public health, including environmental health. He received his training at the University of Missouri at Kansas City, where he received his B.A. degrees in economics and biology as well as his M.D. degree, in 1981. He earned his master in public health degree and completed his preventive medicine residency at Johns Hopkins in 1985. Mitchell is founder and president of the Connecticut Coalition for Environmental Justice. His work is focused on environmental justice, asthma, and air pollution.

Diane-Michele Prindeville
Assistant professor of government, New Mexico State University
Dr. Prindeville's teaching and research interests include American Indian politics, Latina/o politics, environmental policy, and community development. Prindeville has received awards from the American Political Science Association and the Southwestern Political Science Association for her research in the field of race and ethnic politics. Recently, she received a research grant from the Center for American Women and Politics to study American Indian women in tribal governments.

Virginia Ashby Sharpe
Project director, Integrity in Science at the Center for Science in the Public Interest, Washington, D.C.
Dr. Sharpe works for Integrity in Science, Center for Science in the Public Interest, a nonprofit consumer-advocacy organization that focuses on nutrition and health, food safety, alcohol policy, and sound science. Formerly the deputy director and associate for Biomedical & Environmental Ethics, the

Hastings Center, New York, Sharpe worked on ethical issues in health care, biotechnology, and the environment. Her areas of interest are environmental philosophy and ethics, including environmental justice and the ethics of wolf restoration, ethical issues in the structure and delivery of health care, and feminist bioethics.

John A. Stewart

Associate professor of sociology, research associate of the Center for Social Research, University of Hartford

Dr. Stewart's areas of research include environmental sociology and the sociology of science. He is the author of the book, *Drifting Continents and Colliding Paradigms*.

Gerald R. Visgilio

Professor of economics, associate director of the Goodwin-Niering Center for Conservation Biology and Environmental Studies, Connecticut College

Dr. Visgilio is a faculty member in the Department of Economics and serves as the academic advisor to the Certificate Program in Environmental Studies. His research and teaching interests include an economic analysis of environmental and natural resource policy, environmental law, environmental justice, and antitrust law and policy.

Harvey L. White

Associate professor of public and international affairs, University of Pittsburgh

Dr. White has served as director of the university's graduate programs in planning and public administration and currently coordinates undergraduate studies in public administration. He currently serves as chair of the International Conference on Public Management, Policy, and Development and is past president of the Conference of Minority Public Administration. White's current teaching and research interests include the areas of environmental policy and management as well as sustainable development.

Diana M. Whitelaw

Assistant director, the Goodwin-Niering Center for Conservation Biology and Environmental Studies, Connecticut College

Dr. Whitelaw currently coordinates the Certificate Program in Environmental Studies, an interdisciplinary program designed to enhance the undergraduate experience with a strong concentration on environmental issues. Formerly the state Title I director at the Connecticut State Department of Education, Whitelaw managed the federal educational program for disadvantaged children in local school districts.

James Younger
Director, Office of Civil Rights and Urban Affairs, U.S. Environmental Protection Agency, New England
James Younger has led the EPA New England region's Workgroup and the agency's National Program Review Team on Environmental Justice. He has also led the regional effort on the development of a Diversity Action Plan. Younger has an extensive background in administering affirmative action and equal employment opportunity programs, job training programs, and diversity initiatives.